LIVING INTERACTION DESIGN

Bringing together interaction design and the emerging field of biodesign, this book explores the use of living and biofabricated materials in interactive artifacts. As future products become increasingly interactive and complex, careful consideration of their environmental and ethical impacts is more important than ever. While the extraction, production and disposal of materials such as plastics and metals raise serious concerns, biological systems based on living materials offer a more promising path towards regenerative design.

The book introduces the concept of Designed Interactive Living Systems (DILS), a new design paradigm in which living materials actively participate in interactions, facilitating a transition from human-centered to multispecies and more-than-human perspectives. It begins with an overview of the life cycle impacts of interactive products, covering critical phases such as raw materials, manufacturing, use, waste and the environmental, health and social impacts of materials. A wide range of illustrated case studies present solutions to these challenges, showcasing organic electronics, living technologies and living interactive products alongside analyses of existing prototypes and investigations. Finally, the book offers a series of design maps to guide future DILS developments, covering topics such as designing with microorganisms, human–non-human interaction and the role of biolabs as hybrid spaces for experimentation.

Through the presentation of original research and diverse case studies, the book inspires students and designers alike to expand their knowledge and practice. It emphasizes the urgent need to transition from traditional design to biodesign and demonstrates how designers can, and must, take on this role to help shape regenerative, symbiotic futures.

Valentina Rognoli is associate professor at the Design Department of Politecnico di Milano, Italy, and founder of the Materials Design for Transition research group.

Elena Albergati is currently a PhD candidate in Design at the Politecnico di Milano, Italy, involved in research initiatives exploring the role of DILS.

Barbara Pollini is an Italian designer, researcher and educator who is currently assistant professor of Biodesign Integration at Aalto University, Finland.

Carla Langella is architect and associate professor of Industrial Design at the Department of Architecture (DiARC), University of Naples "Federico II", Italy.

LIVING INTERACTION DESIGN

THE TRANSITION TOWARDS DESIGNED INTERACTIVE LIVING SYSTEMS FOR REGENERATIVE ECOLOGIES

VALENTINA ROGNOLI, ELENA ALBERGATI, BARBARA POLLINI, AND CARLA LANGELLA

Routledge
Taylor & Francis Group
LONDON AND NEW YORK

Designed cover image: Getty Images

First published 2026
by Routledge
4 Park Square, Milton Park, Abingdon, Oxon OX14 4RN

and by Routledge
605 Third Avenue, New York, NY 10158

Routledge is an imprint of the Taylor & Francis Group, an informa business

British Library Cataloguing-in-Publication Data
A catalogue record for this book is available from the British Library

ISBN: 978-1-032-43330-1 (hbk)
ISBN: 978-1-032-43328-8 (pbk)
ISBN: 978-1-003-36679-9 (ebk)

DOI: 10.4324/9781003366799

Typeset in Akzidenz Grotesk and Franklin Gothic
by Deanta Global Publishing Services, Chennai, India

CONTENTS

P 198 **CHAPTER 4**

TOWARDS DILS: MAPPING INTERACTIONS WITH BIO-UX

P 234 **CHAPTER 5**

BIOLABS: CROSSROADS OF DESIGN, SCIENCE, COMPUTING AND SUSTAINABILITY FOR DILS

ABOUT THE AUTHORS

Valentina Rognoli is associate professor at the Design Department of Politecnico di Milano, Italy, and founder of the Materials Design for Transition research group. Internationally recognized as a pioneer in materials design, she has been shaping the field for over 25 years through a transdisciplinary approach that bridges design, engineering, life sciences and the humanities.

Her research is driven by a regenerative vision of materials as active agents in rethinking production and consumption systems in harmony with ecosystems. Her work spans biodesign, DIY and circular materials, materials from waste, biofabrication, and interactive and speculative approaches to materiality.

Elena Albergati is currently a PhD candidate in Design at the Politecnico di Milano, Italy, involved in research initiatives exploring the role of Designed Interactive Living Systems (DILS). She previously worked at Mogu, the first Italian company to produce fully circular industrial products based on mycelium, where she gained hands-on experience in biofabrication, sustainable materials and the application of living systems in product design. She coordinated the MOOC Materials Design for Transition towards Regenerative Ecologies and has been actively involved in several Master's programs, organizing workshops on the theory and practice of DILS.

Barbara Pollini is Italian designer, researcher and educator who is currently assistant professor of Biodesign Integration at Aalto University, Finland. She holds a Master's degree in Ecodesign and Eco-innovation and a Master's degree in Computational Design. She was a senior lecturer at Naba Design University in Milan for 12 years, teaching sustainable design and materials, DIY materials, biodesign and design fiction for sustainable futures. In 2023 she completed a PhD in Design at the Politecnico di Milano with research on the potential of biodesign for conventional and regenerative sustainability.

Carla Langella is architect and associate professor of Industrial Design at the Department of Architecture (DiARC), University of Naples "Federico II", Italy. She currently teaches User Experience Design, Biomimetics, and Industrial Design. Since 2006, she has coordinated the Hybrid Design Lab, a research laboratory focused on the interdisciplinary relationships between design and the life sciences, with particular emphasis on biomimetics, design-led materials, design for health, design for heritage, and science communication. She is the author of over 100 scholarly contributions, including peer-reviewed journal articles, book chapters, conference proceedings and monographs.

INTRODUCTION

DOI: 10.4324/9781003366799-1

In light of an escalating environmental crisis and the pervasiveness of digital technologies in our lives, it has become clear that the scope of design must expand – beyond the human, beyond the screen and beyond the illusion of digital immateriality. As designers, we are trained to give form to experience, to create products, interfaces and systems that mediate between people and the world. But today, this mediation must also take into account the material, the environmental and the more-than-human dimensions of our work.

In the field of interaction design in particular, digital and interactive devices, often celebrated for their seamless functionality and abstraction, are deeply rooted in extractive processes and finite and limitated material realities. From the mining of rare earth elements to the energy consumption and the accumulation of electronic waste, the material supports of these technologies reveal a planetary cost that challenges our understanding of innovation and progress. The idea of the "immaterial" digital interface is collapsing under the weight of its environmental footprint.

This book starts from the need to face this urgency – but it does not stop at criticism. Instead, it aims to offer a new vision for the future of design, one that shifts the focus from problems to potential. We propose that meaningful responses to today's environmental and technological challenges require not only new materials and tools, but also a fundamental redefinition of the role and vocabulary of design. This necessitates a profound transformation of both the practice and language of design, including a reconfiguration of the designer's skill set. In this evolving landscape, the designer becomes a biodesigner – a figure capable of recognizing and harnessing the generative potential of hybridization with biology and the vital processes of living organisms. Within this framework, the book specifically explores how such a paradigm shift can unfold within the field of interaction design.

We have chosen to refer to the interactive living systems described in this book as Designed Interactive Living Systems (DILS) to emphasize the specificity of their design origin and the creative intentionality that defines them. The key difference lies not only in the presence of living organisms or in interactivity itself, but in the fact that these systems are conceived and developed by designers – and more specifically, by biodesigners: figures capable of transcending traditional disciplinary boundaries to explore new alliances with living organisms. It is precisely through the vision of the biodesigner that these systems acquire a cultural, critical and innovative value that distinguishes them from non-designed biological practices that lack a conscious design culture.

In the case studies presented in this volume, it is the biodesigner who breaks with the conventions of interaction design as a discipline limited to screens, circuits and plastic, and instead redefines it as a living practice – dynamic, collaborative and in dialogue with living organisms. This perspective brings to light the communicative and interactive potential of living materials such as algae, bacteria, fungi or cellulose, opening up new possibilities for the development of meaningful, sustainable alternatives to conventional digital technologies.

The very notion of interaction is expanded: no longer just between user and product, but between species, between systems and across temporalities. We are invited to design not only for human users, but for coexistence – with other forms of life, seen not as tools, but as co-designers. In this view, life is not a metaphor, but material, presence, voice.

The aim of this book is to map the current state of interaction design's increasing hybridization with biology, and to explore how biobased, biofabricated and living materials can serve not only as sustainable alternatives, but also as regenerative agents capable of profoundly reconfiguring our relationship with technology, matter and life itself.

In doing so, we recognize that the change required is not only technological – it is also methodological, philosophical and ethical. It requires the integration of perspectives from microbiology, synthetic biology, environmental science and posthuman theory, and the adoption of hybrid practices in contexts such as biolabs and biofoundries. It requires new tools, new metrics, new narratives – and, perhaps most importantly, new forms of care.

Through critical reflections, case studies and speculative scenarios, this book outlines an evolving design landscape that values temporality, uncertainty, relationality and mutualism. It is not a handbook, but a terrain: a space to be traversed, where future designers can explore the boundaries of interactivity through the lens of living systems.

As the boundaries between the digital and the biological begin to dissolve, designers are uniquely positioned to lead this transition – not as passive adopters of innovation, but as active agents of transformation. We need to rethink what we make, how we make it and with whom we make it.

This book is both an invitation and a provocation – to design more critically, to act more deeply, and to imagine more radically. The future of interaction design is not just digital. It is alive.

CHAPTER 1

ENVIRONMENTAL IMPACT OF ELECTRONIC AND INTERACTIVE DEVICES

DOI: 10.4324/9781003366799-2

AN IMPACTING LIFECYCLE

When discussing *digital technology*, we refer to systems and processes that use binary language (sequences of 0s and 1s) to represent, process, store and transmit information. Digital technologies are often perceived as ephemeral and intangible because of their ability to create experiences and environments that seem detached from tangible reality. Technologies such as cloud computing, social media and virtual reality develop platforms that project users into a "digital space" without physical boundaries and disconnected from materiality. Many functions once tied to tangible objects, such as CDs, books or maps, have been dematerialized and transferred to digital formats, reinforcing the perception of intangibility. Instant and ubiquitous access to data feeds the idea that the digital is "everywhere", an omnipresent element that transcends physicality (Abowd & Schilit, 1997; Bolter & Grusin, 2022).

This perception of ephemerality has been reinforced by *artificial intelligence* (AI). Sophisticated algorithms and advanced models operate behind the scenes, making digital experiences even more seamless and seemingly devoid of visible infrastructures. With its ability to analyze, predict and adapt in real time, AI reinforces the illusion of a self-contained, immaterial virtual world, fostering the notion that digital technology is an ethereal, almost omnipresent entity.

We thus find ourselves immersed in an intangible realm where the connection to the physical world, our bodies, the environment and direct interactions with others seem to fade. This ephemeral dimension of the digital has been powerfully imagined in visionary and speculative works such as the binary language of *The Matrix* (1999) or the on-air displays of *Minority Report* (2002). Similarly, some episodes of *Black Mirror* (2011–today), the immersive virtual world of *Ready Player One* (2018) or the existential simulation of

The Thirteenth Floor (1999) emphasize the illusion of an intangible realm suspended between reality and the virtual.

Although digital technologies are often perceived as ephemeral and intangible, this perception is only partially accurate. Digital technologies always have a physical presence, as tangible interaction research has long emphasized (Ishii et al., 2004). They evoke an intangible realm where the connection to the physical world seems to disappear, but they remain deeply connected to material devices, physical infrastructures and environmental resources, making their impact far more tangible than it might initially appear. Digital technologies require infrastructures such as servers, data centers, hardware devices, network cables and sensors. Their production and operation depend on material resources such as rare minerals, energy and complex industrial processes.

Electronic and interactive devices are the physical means through which digital technologies manifest themselves, interact with users and influence the real world. These devices combine hardware and software components to enable functionalities that facilitate human interaction with technology and the environment. Devices such as smartphones, computers and sensors host and process digital technologies, while electronic circuits and microprocessors provide the physical platform necessary to run complex software. These technologies enable devices to respond to user input in real time through interfaces such as touch screens, voice commands or gesture recognition and to significantly enhance the user experience through artificial intelligence and machine learning (ITU, 2024).

Digital connectivity allows these devices to remain constantly connected via Wi-Fi, 5G or other technologies, enabling seamless communication and data exchange between systems. Many devices provide physical feedback, such as vibrations or sounds, and visual feedback, such as LED signals or displays, creating more immersive and intuitive user experiences. This symbiotic relationship between digital technologies and electronic devices drives continuous innovation and underscores how the digital and physical worlds have become intertwined, fostering an ever-evolving technological ecosystem.

The lifecycle of an electronic and interactive device – from production to use, obsolescence and disposal – illustrates the tangible foundations and significant impact of digital technologies. These devices have become essential tools for communication, work, entertainment and countless other activities. However, their lifecycle is driven by the extraction and consumption of finite resources, and each device has a significant environmental footprint. Although digital technologies may seem intangible, creating and maintaining the software and data that power them require considerable energy, materials and physical space, particularly for server storage and data transmission. Energy demands extend beyond personal devices to the vast infrastructure of data centers, whose power consumption is critical to the digital ecosystem. Our interactions with digital technologies – such as browsing, streaming or other online behaviors – contribute directly to energy consumption, linking digital activities to broader environmental impacts. (Cara & Palazzo, 2022). This cycle culminates in generating electronic waste (e-waste), exacerbated by the relentless pursuit of newer digital technologies and devices. This trend is often driven by functional advancement and aesthetic obsolescence, accelerating the replacement of devices long before their functional end of life and amplifying the tangible impacts of our digital age.

This holistic view emphasizes the profound physicality of digital technology, challenges the misconception of its immateriality and highlights the urgent need for sustainable practices, innovative materials and robust policy frameworks to mitigate the environmental impact of digital systems and devices (Shirvani & Ghasemshirazi, 2024; Garrett-Peltier, 2017).

This is also because our human existence is increasingly surrounded by and dependent on electronic devices. These devices, which can connect people and facilitate interaction and communication, are constantly evolving and being introduced in various sectors such as healthcare, security, telecommunications, entertainment and others. They have found their way into almost every home and pocket, their usefulness and versatility making them indispensable. Devices ranging from the simplest, such as televisions and radios, to the most intimate, such as mobile phones, to the most sophisticated, such as digital cameras, medical equipment, computers and even cars, among others, have become part of our daily lives. Although electronics offer many benefits, they are not without their drawbacks.

This discussion will not delve into the opportunities or positive and negative consequences of the proliferation of electronic devices, leaving that to researchers in sociology, ethics, psychology and anthropology who have long studied these issues (Tran et al., 2020; Meyrowitz, 1986; Ray & Jat, 2010). Instead, the aim is to emphasize that these devices, which we all buy, use, dispose of and replace, have an environmental impact linked to their design and usage patterns. The consequences of their design, production and consumption practices are often overlooked, resulting in a general lack of awareness of the real impact of electronic devices (Clark & Clark, 2019; Parajuly et al., 2019).

Electronic tangible media allow users to interact with them through various interfaces, including touch screens, buttons, various physical components and even voice commands or gestures. This notion aligns with early visions of tangible interaction that aimed to bridge the gap between the physical and digital worlds (Ishii & Ullmer, 1997). These devices typically incorporate computing functionality and are designed to provide real-time interaction, allowing users to enter commands or manipulate data and receive immediate feedback. Because of their ability to facilitate user interaction, these devices have become versatile tools for communication, productivity, entertainment and information dissemination. Examples include smartphones, tablets, laptops, personal computers, game consoles, smart TVs, wearable devices and virtual reality (VR) and augmented reality (AR) devices. The digital technology market is constantly evolving with the introduction of new physical devices.

The electrical and electronic industry has made significant economic and social contributions worldwide for almost a century. However, electronic device production, use and disposal are resource-intensive processes with profound environmental and climate impacts (Rene et al., 2021; Ismail & Hanafiah, 2020). The magnitude of these impacts depends on consumption patterns and product lifetimes. In 1999, sociologist Neil Gross envisioned a future in which almost everything is connected to the Internet:

> In the next century, planet Earth will put on an electronic skin. It will use the Internet as a scaffold to support and transmit its sensations. This skin is already being stitched together. It consists of millions of embedded electronic measuring devices: thermostats, pressure gauges, pollution detectors, cameras, microphones, glucose sensors, ECGs and electroencephalographs. These will probe and monitor cities and endangered species, the atmosphere, our ships, highways and truck fleets, our conversations, our bodies – even our dreams.
>
> (Gross, 1999; Resch et al., 2012)

The vision was that the Internet of Things (IoT) would contribute to a greener world, with sensors detecting pollution, smart thermostats saving energy and agricultural technology conserving water. However, the vast network of devices could also hurt the environment (Finley, 2014). A primary concern at the moment is the environmental impact of wearable devices. These devices raise environmental and human health concerns throughout their lifecycle.

Electrical and electronic equipment can contribute to environmental pollution during manufacturing (Akram et al., 2019). The production of electronic equipment requires the extraction of raw materials such as metals and minerals, which can lead to habitat destruction, soil erosion and water pollution. In addition, greenhouse gases and toxic chemicals are released into the environment during production.

The extraction and use of these materials, including the mining and processing of copper, critical raw materials (CRMs), rare earth elements (REEs) and metals such as tantalum, plastics and glass, has significant environmental and social impacts, contributing to resource depletion and the use of chemicals and water.

At the end of their lifecycle, electronic devices become electronic waste, or e-waste, one of the fastest-growing waste streams in the world, increasing by 4% per year (Baldé et al., 2017). E-waste contains hazardous substances such as lead, mercury, cadmium and flame retardants. If not properly managed, these substances can contaminate soil, water and air, exacerbating pollution (He et al., 2022; Weerasundara et al., 2020; Verma, 2020). The environmental sustainability of electronic equipment depends on several factors, including the materials used in its manufacture, its energy efficiency and its ability to be repaired, expanded and upgraded. Some electronic devices, such as smartphones and tablets, have a relatively high environmental footprint due to the energy and material resources required for their production and frequent replacement. Strategies are being considered to reduce the environmental impact of these devices, including adopting more energy-efficient manufacturing processes, using recycled materials and promoting responsible recycling and disposal practices (Cenci et al., 2022). However, these measures alone are not sufficient to initiate a new paradigm of interaction design that is more in line with nature's approach to its interactive artifacts, such as the skin of organisms, the antennae of insects or the roots of plants (Höök & Löwgren, 2021; Kulkarni & Saraf, 2019).

The current environmental impact of electronic devices requires a comprehensive investigation to understand better the possibilities for modification and redesign and the exploration of new material possibilities. This examination must cover the entire lifecycle of electronic equipment. It requires a rigorous environmental assessment that evaluates their impact from the extraction of raw materials through production, transport and use to disposal or recycling at the end of their lifecycle (Subramanian & Yung, 2016; Andrae, 2015; Cenci et al., 2022).

This holistic approach is essential to inform decisions and strategies to minimize the environmental footprint of electronic equipment and promote sustainability in its design and use. By examining the environmental impacts associated with raw material sourcing, manufacturing processes, transport and end life, we can identify areas for optimization and greening.

The different stages in the lifecycle of interactive electronic equipment are now examined, highlighting their various impacts on the environment, human health and society.

THE HIDDEN COSTS: RAW MATERIALS

The production of interactive and electronic devices relies on a wide range of raw materials, including precious and rare metals. Each component must be extracted and processed, which poses significant environmental, political and socio-economic challenges. These include habitat destruction, pollution and conflicts over resources.

According to Statista,[1] this represents approximately 86.11% of the global population – a dramatic increase from 2016, when only

366.8 million people, or 49.40% of the worldwide population, owned smartphones. The rapid adoption of smartphones has led to an exponential increase in demand. Between 2017 and 2022, the number of smartphone users is expected to grow by 49.89%, and global smartphone subscriptions are forecast to reach 7.8 billion by 2028, with China, India and the United States leading the way.

With the global mobile user population forecast to reach 7.16 billion by 2026 and 7.516 billion within the following three years, demand for the raw materials needed to manufacture these devices will continue to soar. Smartphones, a ubiquitous example of electronic devices that connect and facilitate interaction, illustrate this dependency. These devices, particularly those with aluminum bodies, are made up of a wide range of materials, including silicon for microchips, lithium for batteries and traces of precious metals such as gold and platinum in the circuitry. Each step in extracting and processing these materials has a profound environmental and social impact, highlighting the urgent need for sustainable practices in producing and disposing of electronic equipment.

It's important to note that tangible media extends far beyond smartphones, further increasing the demand for materials. The production of electronic devices relies heavily on various metals, including precious and rare metals. Research from Yale University highlights supply constraints for several metals critical to the rapidly growing electronics sector, such as gallium and selenium, mainly due to supply-related risks (Graedel et al., 2015). Metals such as platinum, gold and mercury face serious environmental challenges during mining and processing, while others such as chromium, niobium, tungsten and molybdenum face supply vulnerabilities (Raabe, 2023).

Several factors exacerbate these problems. Scarcity and geopolitical tensions play a key role, as metals such as REEs, gold and platinum are not only finite but also concentrated in certain regions. For example, China's dominance in the supply of rare earth elements, essential for many electronic devices, raises concerns about supply chain vulnerability and price volatility (Köhler, 2012).

The environmental impacts of raw materials extraction include deforestation, soil erosion, water pollution and loss of biodiversity. Refining metals and producing plastics for electronic devices further exacerbate energy demand and greenhouse gas emissions. In addition, human rights abuses are closely linked to mining practices. Conflict minerals such as tin, tantalum, tungsten and gold often come from war-torn regions where mining is associated with violence, child labor and other human rights abuses. The lack of traceability in the supply chain complicates efforts to ensure ethical sourcing and avoid inadvertently supporting harmful practices.

Interactive and electronic devices can contain up to 60 elements from the periodic table (Eionet Report – ETC/WMGE 2020/3). To comprehensively address these environmental and societal challenges, examining the specific impacts of key materials commonly found in electronic devices, such as aluminium, gold, silver, cobalt, copper, lead, nickel, beryllium and plastics, is critical. Each material contributes uniquely to the environmental footprint of these products, highlighting the importance of sustainable design, manufacturing and recycling practices.

Let us look at each of these in more detail and consider their challenges.

Aluminium, widely used in electronic devices for its strength and heat dissipation properties, has significant environmental impacts resulting from its extraction and production processes. Bauxite mining, its primary source, often leads to deforestation, habitat destruction and loss of biodiversity while also causing soil erosion that undermines agricultural productivity and local ecosystems.

Bauxite processing involves hazardous chemicals such as caustic soda, resulting in severe water pollution threatening aquatic life and human health. Aluminum production is also highly energy intensive, with much of the electricity required coming from non-renewable sources such as coal, contributing significantly to CO2 emissions and global climate change.[2] In addition, the process generates toxic waste that poses risks to both the environment and human health, including respiratory diseases such as pneumoconiosis among workers.

The transition to renewable energy sources in aluminum production and adopting low-emission technologies are critical to reducing its environmental footprint. Continued innovation in production methods can help balance the demand for aluminum with the need to mitigate its environmental and social impacts (Georgitzikis et al., 2021).

Gold and silver are essential in interactive and electronic devices due to their exceptional electrical conductivity, corrosion resistance and alloy compatibility. Gold, with its low electrical resistance and resistance to tarnishing, is widely used in high-end connectors. In contrast, silver, known for its superior electrical and thermal conductivity, is used in switches, relays and electrical contacts (Goodman, 2002). Both metals are often alloyed to enhance specific properties, balancing performance and cost.

However, the extraction of gold and silver has a significant environmental impact (Berthet et al., 2024). Gold mining involves extensive excavation and the use of toxic chemicals such as cyanide and mercury, leading to contamination of rivers and groundwater. This process results in habitat destruction, biodiversity loss, soil erosion and sedimentation in water bodies, adversely affecting ecosystems. Air pollution from blasting and ore processing further degrades air quality (Schön-Blume et al., 2021). Gold mining also exacerbates water scarcity and depletes natural reserves, disrupting the ecological balance, and is often associated with human rights abuses, including child labor and poor working conditions.

Silver mining, while less destructive, poses ecological risks, including the inhibition of microorganisms in wastewater by silver nanoparticles. Rapid resource depletion and rising costs have also made silver less economically viable for specific applications. Efforts to mitigate these impacts include stricter regulations, cleaner technologies and improved practices. However, sustainable approaches remain critical to minimize environmental damage and ensure the responsible use of these finite resources in interactive and electronic devices.

Cobalt is critical for lithium-ion batteries in smartphones, laptops and electric vehicles. However, its extraction poses serious socio-economic and environmental challenges. A significant proportion of cobalt comes from the Democratic Republic of Congo, where mining is associated with child labor, unsafe working conditions and health risks such as respiratory problems (Karduri, 2023). Amnesty International reports have also highlighted the involvement of children in the extraction of cobalt for electronic devices in the DRC, where children can work up to 12 hours a day for meager wages of one to two dollars a day. In 2014, it was reported that around 40,000 children were employed in cobalt mining in the DRC.[3] Environmentally, cobalt mining depletes water and energy resources, releases harmful substances such as sulfur dioxide, and produces toxic waste that contaminates soil and water, threatening ecosystems and human health.

Addressing these impacts requires stronger regulations and sustainable practices.

Copper is essential in electronics, particularly for printed circuit boards (PCBs), due to its excellent electrical conductivity, thermal efficiency and malleability. It is necessary for connecting and powering electronic components.

However, copper mining has a significant environmental impact (Lu et al., 2023). Open pit and underground mining methods often result in habitat destruction, deforestation, soil erosion, water pollution and greenhouse gas emissions. In addition, the mining process requires significant energy and water resources, further increasing the environmental footprint.[4] Addressing these impacts requires more sustainable mining practices and resource-efficient technologies.

Lithium, the lightest metal, is indispensable in modern technologies, particularly lithium-ion batteries used in portable devices, electric vehicles and energy storage systems. Its high energy storage capacity, low density and unique properties also make it valuable in aerospace alloys and medical applications (Shakoor et al., 2023; Krishnan & Gopan, 2024).

However, the extraction of lithium poses significant environmental challenges. The process requires intensive water use, risks contaminating local ecosystems and contributes to greenhouse gas emissions (Kaunda, 2020; Mrozik et al., 2021). Despite these impacts, lithium is central to advancing sustainable energy solutions, particularly in the electrification of transport.

Australia, China and Chile dominate the global lithium market, reflecting the metal's strategic importance as demand for high-tech and renewable energy applications grows. Sustainable mining and processing methods are critical to mitigating environmental impacts while supporting the global energy transition.

Nickel is widely used in electronics for its exceptional electrical conductivity, corrosion resistance and thermal properties. It is critical in rechargeable batteries such as nickel metal hydride

(NiMH) and nickel-cadmium (NiCd). However, its extraction from low-grade ores is energy intensive and relies heavily on fossil fuels, resulting in significant greenhouse gas emissions and a high carbon footprint.

Nickel mining, particularly open-pit mining, causes habitat destruction, biodiversity loss and soil and water contamination (Sonter et al., 2018). The release of nickel compounds into the environment harms ecosystems and human health, with nickel ranked as the seventh most harmful metal for its ecological and health impacts (Genchi et al., 2020). Sustainable mining practices are essential to reduce these risks.

Beryllium is valued in aerospace, defense and electronics for its strength, lightweight and heat resistance. However, it is highly toxic, and prolonged exposure has been linked to chronic beryllium disease (CBD) and cancer (MacMurdo et al., 2020). As a Class A carcinogen, its handling requires strict safety protocols to protect human health.

Plastics are widely used in electronics, with materials such as polycarbonate (PC) and acrylonitrile butadiene styrene (ABS) offering durability and versatility. However, the production of plastics releases harmful chemicals such as benzene and toluene, which contribute to respiratory disease and cancer in nearby communities. Plastics also pollute water sources, affect aquatic life and accumulate in the food chain.

Black plastics, commonly used in electronics, pose recycling challenges because they absorb light, making them difficult to sort in recycling facilities. They also often contain additives such as brominated flame retardants, which pose potential health risks (Alassali et al., 2020). Promoting sustainable plastic alternatives and improving recycling technologies are critical to minimizing these impacts.

IMPACT OF MANUFACTURING

The environmental impact of interactive and electronic devices such as laptops and smartphones, to name a few, goes far beyond their energy consumption, which is often the primary focus when assessing their environmental footprint.

According to a study by the Platform for Accelerating the Circular Economy (PACE) and the World Economic Forum, by 2040, the production of electronic equipment is expected to account for 14% of total greenhouse gas emissions. This highlights the urgent need for sustainable manufacturing practices.[5]

The most significant environmental impact of these appliances occurs during the manufacturing phase. This phase accounts for most energy consumption and CO_2 emissions, particularly in Asia, where fossil fuels dominate energy production. This reliance on non-renewable energy sources significantly contributes to the global carbon footprint.

The global market for electronic devices is growing at a compound annual growth rate (CAGR) of 6.9%, further exacerbating their environmental impact. For example, producing a single laptop in one year can use more than 246 billion liters of water, more than 39,360 tonnes of fuel and more than 3,068 million tonnes of chemicals. Such resource-intensive processes underline the need for innovative solutions to reduce consumption and emissions.[6]

Companies increasingly extract raw materials and refine assembly processes to meet the growing demand for portable computers. Production is often outsourced to countries like China and Malaysia, leveraging low-cost labor. However, this approach raises concerns about these regions' environmental sustainability and social equity.

Manufacturing microchips, vital components in desktop computers, is another major environmental challenge. On average, 132 kilograms of fuel are used to produce these small silicon parts – more than the energy needed to power a laptop for three years.[7] Similarly, producing copper foil for printed circuit boards involves energy-intensive smelting and refining processes that emit greenhouse gases and air pollutants.

In addition, manufacturing generates waste products such as slag, tailings and emissions, highlighting the need for robust waste management and treatment systems to mitigate environmental damage.

The environmental and social impacts of electronics manufacturing are profound, including significant greenhouse gas emissions, resource depletion and the potential destabilization of vulnerable regions. Addressing these challenges will require coordinated efforts by industry, governments and consumers.

IMPACT OF TRANSPORTATION AND DISTRIBUTION

The logistics industry, which plays a vital role in transporting goods, significantly impacts the environment, mainly through air pollution and resource consumption. This problem is exacerbated by the rapid growth of e-commerce and the increasing demand for electronic products (Siragusa & Tumino, 2021).

Globally, the movement of goods is a major contributor to air pollution, emitting greenhouse gases such as carbon dioxide, methane and nitrous oxide, as well as pollutants such as nitrogen oxides, sulfur oxides and particulate matter. These emissions are a critical concern as the logistics sector accounts for around a quarter of global carbon dioxide emissions. Without significant intervention, this figure could rise to 40% by 2050, according to the European Environment Agency.[8]

The carbon footprint of different modes of transport – road, water and air – not only exacerbates global warming but also poses direct health risks from air pollution.

Logistics activities also put severe pressure on global water resources. Maritime transport, for example, pollutes aquatic ecosystems and marine life by discharging wastewater, solid waste and hazardous substances such as oil and chemicals. Underwater noise pollution from shipping disrupts marine species that rely on sound for communication, navigation and feeding, causing disorientation, stress, and even stranding. In addition, oil spills, whether accidental or due to improper disposal, cause long-term damage to marine ecosystems and coastlines.

Another challenge is the logistics sector's heavy reliance on non-renewable fuels, particularly in large-scale operations involving cargo aircraft and container ships. While there has been some progress in introducing electric vehicles, decarbonizing the entire transport industry remains formidable (Gray et al., 2021).

The transport of electronic goods, including portable and fixed equipment, requires extensive packaging to ensure product protection and effective marketing. However, improper disposal of this packaging exacerbates landfill problems. Certain materials, particularly plastics and composites, present significant recycling challenges, complicating waste management efforts and underscoring the urgent need for more sustainable packaging solutions in the interactive and electronic device sector.

IMPACT OF USE

Interactive and electronic devices have a significant environmental impact during their use phase, mainly due to energy consumption, including standby energy consumption (Hischier et al., 2020). Energy consumption varies depending on the type of device, its features, user behavior and energy efficiency. To mitigate this impact, users must adopt responsible habits such as adjusting screen brightness, enabling low-power modes, and turning off devices when not in use. These small changes can collectively reduce the environmental footprint of electronic devices.

While it is impractical to quantify the energy consumption of all electronic devices, raising user awareness is crucial. Incorporating energy-aware technologies and innovative design features can play a key role in encouraging sustainable choices. For example, interactive energy dashboards and green feedback systems, such as smart thermostats and lighting controls, provide real-time feedback on energy consumption. These tools often use visual cues, gamification or incentives to encourage users to adopt energy-saving behaviors (Chalal et al., 2022).

Despite progress, the industry faces ongoing challenges related to energy consumption. Continued demand for chargers and batteries, particularly in developed regions, exacerbates energy inefficiencies. Households often accumulate unused chargers, many of which remain plugged in, consuming power even when not actively charging devices. This phenomenon, commonly referred to as standby, phantom or vampire power, is the energy consumed by devices and chargers even when switched off or in standby mode. While the energy consumption of an individual device may seem small, it becomes significant when aggregated across millions of households (Mullai & Sivasamy, 2017).

Addressing these issues requires a multifaceted approach to bridge the gap between user habits, device design and sustainable energy practices.

IMPACT OF REPAIR/RECYCLE/DISPOSAL

The end-of-life management of interactive and electronic devices is a crucial challenge to reduce their environmental impact. Proper repair, recycling and disposal are key practices to mitigate negative impacts, limit the generation of electronic waste (e-waste) and preserve valuable resources. E-waste includes discarded electronic and interactive devices that are technologically and aesthetically obsolete and are one of the fastest-growing categories of waste worldwide. Among these, Printed Circuit Boards (PCBs) represent a particularly challenging component due to their complex material composition and recycling processes (Selvakumar et al., 2023). According to the United Nations, e-waste is defined as "any discarded product with a battery or plug containing toxic and hazardous substances, such as mercury, that may pose a serious risk to human and environmental health" (Heacock et al., 2016; Juma Nyeko et al., 2023).

Rapid technological advances have exacerbated this crisis, shortening product lifespan and increasing device replacement rates. By 2022, it is estimated that 62 million tonnes of e-waste will be generated globally, an increase of 82% since 2010. However, only 22% of this waste was formally collected and recycled, highlighting significant gaps in global management practices.[9] Most waste ends up in landfills, incinerators or improperly accumulated, causing severe environmental and public health damage (Fawole et al., 2023). According to the Global E-waste Monitor 2024,[10] e-waste is increasing by 2.6 million tonnes per year and is expected to reach 82 million tonnes by 2030, a further 33% increase from 2022 levels.

In Europe, despite a recycling rate of 54%, large-scale adoption of circular economy practices is hampered by an underdeveloped repair culture and low consumer awareness. Repair, crucial to extending the life of devices, is often neglected due to structural and cultural barriers: limited access to spare parts, high costs and lack of adequate technical support. Many consumers find it easier to discard and replace products rather than repair them, highlighting the need for support systems and training to incentivize repair and reuse (Liu et al., 2023; Van den Berge et al., 2023; McCollough, 2009; Bovea et al., 2020). To address these issues, the European Union has introduced innovative legislation, such as the *Right to Repair*,[11] to make repair more accessible and affordable. The new EU directive, to be fully adopted by 2026, includes extended warranties for repaired products, easier access to spare parts and awareness-raising campaigns to educate consumers and promote more sustainable choices[12] (Nagase & Uehara, 2024). Following the example of the French repairability index, the EU is considering the introduction of standardized rating systems to help consumers consider the repairability of products at the time of purchase.

In parallel, the *Waste Electrical and Electronic Equipment* (WEEE) directive established a comprehensive framework for the management of e-waste in the EU. The directive sets specific targets for collecting, recycling and recovering e-waste and encourages producers to design equipment that is easier to repair, recycle and disassemble. It also requires EU Member States to set up accessible collection points and adequate recycling facilities. Based on the principle of Extended Producer Responsibility (EPR), the directive makes producers responsible for the end-of-life management of their products. It encourages them to design environmentally friendly equipment prioritizing durability, repairability and recyclability (European Commission, 2024).

Globally, the situation is even more complex. In many regions, e-waste is recycled through informal practices that pose serious health and environmental risks. Hazardous substances such as lead, mercury and cadmium contaminate soil and water, causing long-term damage. Promoting a circular economy based on durability, repairability, and recyclability is essential to mitigate these impacts (Castle-Green & Sailaja, 2024). Initiatives such as Eco-design regulation[13] and the right to repair are gaining ground, particularly in the EU, in line with the objectives of the European Green Deal. The recycling of e-waste not only enables resource recovery and reduces environmental impacts but also generates economic benefits by reducing the demand for virgin materials and creating jobs in the recycling industry (Gamage et al., 2023).

Finally, the use of biologically inspired materials and mechanisms is emerging as a complementary solution to reduce the environmental impact of interactive technologies. Advances in synthetic biology offer new opportunities for designing living and sustainable technologies, promoting a more responsible approach to developing interactive and electronic devices.

NOTES

1 Number of smartphone users worldwide from 2014 to 2029: https://www.statista.com/forecasts/1143723/smartphone-users-in-the-world
2 UK Green Building Council. *Aluminium: Embodied Ecological Impacts*. https://ukgbc.org/our-work/topics/embodied-ecological-impacts/aluminium/. Accessed 30 March 2025.
3 "'This is what we die for': Human Rights Abuses in the Democratic Republic of the Congo Power the Global Trade in Cobalt". Retrieved March, 2025, from https://www.amnesty.org/en/documents/afr62/3183/2016/en/
4 "Top 10 Environmental Challenges in Copper Mining", available at: https://medium.com/cryptal-global/top-10-environmental-challenges-in-copper-mining-5e09a58f1672
5 "A New Circular Vision for Electronics: Time for a Global Reboot". Retrieved March, 2025, from https://www3.weforum.org/docs/WEF_A_New_Circular_Vision_for_Electronics.pdf
6 "The Environment and Electronic Devices". Retrieved March, 2025, from https://www.tecnologialibredeconflicto.org/en/environment/
7 "Behind the Figures: Understanding the Environmental Impacts of ICT and Taking Action". Retrieved March, 2025, from https://www.greens

-efa.eu/files/assets/docs/ict_environmental_impacts-behind_the_figures-5low.pdf

8 "Transport and Mobility". Retrieved March, 2025, from https://www.eea.europa.eu/en/topics/in-depth/transport-and-mobility

9 Andrew Whalen, "Your Drawer Full of Old Cables Is Worth More Than You Think," *Vox*, 13 March 2024. Retrieved March, 2025, from https://www.vox.com/technology/389775/ewaste-electronics-recycling-tech.

10 United Nations Institute for Training and Research (UNITAR), "The Global E-waste Monitor 2024 – Electronic Waste Rising Five Times Faster than Documented E-waste Recycling", 2024. Retrieved March, 2025, from https://ewastemonitor.info/

11 "Directive on Repair of Goods – Promoting Repair Contributes to Sustainable Consumption". Retrieved March, 2025, from https://commission.europa.eu/law/law-topic/consumer-protection-law/directive-repair-goods_en

12 "New EU Rules Encouraging Consumers to Repair Devices Over Replacing Them". Retrieved March, 2025, from https://www.europarl.europa.eu/news/en/press-room/20231117IPR12211/new-eu-rules-encouraging-consumers-to-repair-devices-over-replacing-them

13 "Ecodesign for Sustainable Products Regulation – Making Sustainable Products the Norm in the EU" (entered into force on 18 July 2024). Retrieved March, 2025, from https://commission.europa.eu/energy-climate-change-environment/standards-tools-and-labels/products-labelling-rules-and-requirements/ecodesign-sustainable-products-regulation_en

REFERENCES

Abowd, G. D., & Schilit, B. N. (1997). Ubiquitous computing: The impact on future interaction paradigms and HCI research. In *CHI '97 extended abstracts on human factors in computing systems* (CHI EA' 97) (pp. 221–222). Association for Computing Machinery. https://doi.org/10.1145/1120212.1120354

Akram, R., Natasha, Fahad, S., Gulzar, A., Shahzad, U., & Duan, Y. (2019). Trends of electronic waste pollution and its impact on the global environment and ecosystem. *Environmental Science and Pollution Research*, *26*, 16923–16938. https://doi.org/10.1007/s11356-019-04998-2

Alassali, A., Barouta, D., Tirion, H., Moldt, Y., & Kuchta, K. (2020). Towards a high-quality recycling of plastics from waste electrical and electronic equipment through separation of contaminated fractions. *Journal of Hazardous Materials*, *387*, 121741. https://doi.org/10.1016/j.jhazmat.2019.121741

Andrae, A. S. (2015). Lifecycle assessment of consumer electronics: A review of methodological approaches. *IEEE Consumer Electronics Magazine*, 5(1), 51–60.

Baldé, C. -P., Forti, V., Gray, V., Kuehr, R., & Stegmann, P. (2017). The Global E-waste Monitor – 2017, United Nations University (UNU), International Telecommunication Union (ITU), & International Solid Waste Association (ISWA), Bonn/Geneva/Vienna. Retrieved March, 2025, from https://collections.unu.edu/eserv/UNU:6341/Global-E-waste_Monitor_2017__electronic_single_pages_.pdf

Berthet, E., Lavalley, J., Anquetil-Deck, C., Ballesteros, F., Stadler, K., Soytas, U., Hauschild, M., & Laurent, A. (2024). Assessing the social and environmental impacts of critical mineral supply chains for the energy transition in Europe. *Global Environmental Change*, *86*, 102841. https://doi.org/10.1016/j.gloenvcha.2024.102841

Bolter, J. D., & Grusin, R. (2022). *Remediation: Understanding new media* (Rev. ed.). MIT Press.

Bovea, M. D., Ibáñez-Forés, V., & Pérez-Belis, V. (2020). Repair vs replacement: Selection of the best end-of-life scenario for small household electric and electronic equipment based on life cycle assessment. *Journal of Environmental Management*, *254*, Article 109679. https://doi.org/10.1016/j.jenvman.2019.109679

Cara, F., & Palazzo, G., (2022). With or without you. Gli impatti sull'ambiente e la società dell'infrastruttura digitale. In AA.VV (Ed.), *Ecologia Digitale. Per una tecnologia al servizio delle persone, società e ambiente* (pp. 37–47). Altraeconomia, Milano.

Castle-Green, T., & Sailaja, N. (2024). Addressing E-waste: Repair Café processes as barriers to repair of smart devices. *Base Diseño E Innovación*, *9*(10), 150–161. https://doi.org/10.52611/bdi.num10.2024.999

Cenci, M. P., Scarazzato, T., Munchen, D. D., Dartora, P. C., Veit, H. M., Bernardes, A. M., & Dias, P. R. (2022). Eco-friendly electronics–a comprehensive review. *Advanced Materials & Technologies*, *7*, 2001263. https://doi.org/10.1002/admt.202001263

Chalal, M. L., Medjdoub, B., Bezai, N., Bull, R., & Zune, M. (2022). Visualisation in energy eco-feedback systems: A systematic review of good practice. *Renewable and Sustainable Energy Reviews*, *162*, 112447. https://doi.org/10.1016/j.rser.2022.112447

Clark, L. A., & Clark, W. R. (2019). Global E-waste: Unintended consequences of marketing strategies necessitates a plan for change. *Journal of Managerial Issues*, *31*(3), 331–347. http://www.jstor.org/stable/45206628

European Commission. (2024). *Waste Electrical and Electronic Equipment (WEEE).* Retrieved March, 2025, from https://environment.ec.europa.eu/topics/waste-and-recycling/waste-electrical-and-electronic-equipment-weee_en

Fawole, A. A., Orikpete, O. F., Ehiobu, N. N., Oloyede, O. E., Adeyanju, O. D., Adegbite, K. A., … Adewumi, A. (2023). Climate change implications of electronic waste: Strategies for sustainable management. *Bulletin of the National Research Centre*, *47*, 147. https://doi.org/10.1186/s42269-023-01124-8

Finley, K. (2014). The internet of things could drown our environment in gadgets. *Wired*. Retrieved March, 2025, from https://www.wired.com/2014/06/green-iot/

Gamage, L. E. N., Basak, A. K., Pramanik, A., Karunasena, G. G. A. R., & Islam, M. R. (2023). Recycling of gold and silver from electronic waste–a review. *Materials Circularity Economy*, *5*, 8. https://doi.org/10.1007/s42824-023-00081-x

Garrett-Peltier, H. (2017). Green versus brown: Comparing the employment impacts of energy efficiency, renewable energy, and fossil fuels using an input-output model. *Economic Modelling*, *61*, 439–447. https://doi.org/10.1016/j.econmod.2016.11.012

Genchi, G., Carocci, A., Lauria, G., Sinicropi, M. S., & Catalano, A. (2020). Nickel: Human health and environmental toxicology. *International Journal of Environmental Research and Public Health*, *17*(3), 679. https://doi.org/10.3390/ijerph17030679

Georgitzikis, K., Mancini, L., d'Elia, E., & Vidal-Legaz, B. (2021). *Sustainability aspects of bauxite and aluminium: Climate change, environmental, socio-economic and circular economy considerations*. Publications Office of the European Union. Retrieved March, 2025, from https://rmis.jrc.ec.europa.eu/uploads/library/jrc125390_sustainability_profile_bauxite__aluminium_online.pdf

Goodman, P. (2002). Current and future uses of gold in electronics. *Gold Bulletin*, *35*(2), 21–26. https://doi.org/10.1007/BF03214833

Graedel, T., Harper, E., Nassar, N., Nuss, P., & Reck, B. (2015). Criticality of metals and metalloids. *Proceedings of the National Academy of Sciences of the United States of America*, *112*(14), 4257–4262. http://doi.org/10.1073/pnas.1500415112

Gray, N., McDonagh, S., O'Shea, R., Smyth, B., & Murphy, J. D. (2021). Decarbonising ships, planes and trucks: An analysis of suitable low-carbon fuels for the maritime, aviation and haulage sectors. *Advances in Applied Energy*, *1*, 100008. https://doi.org/10.1016/j.adapen.2021.100008

Gross, N. (1999). The earth will don an electronic skin. *BusinessWeek Online*. Retrieved March, 2025, from https://www.bloomberg.com/news/articles/1999-08-29/14-the-earth-will-don-an-electronic-skin

He, Z., Yue, Y., & Wang, Y. (2022, April). The hazards, treatment measures, and sustainable development of electronic waste. In *IOP conference series: Earth and environmental science* (Vol. 1011, No. 1, p. 012023). IOP Publishing.

Heacock, M., Kelly, C. B., Asante, K. A., Birnbaum, L. S., Bergman, Å. L., Bruné, M. N., Buka, I., Carpenter, D. O., Chen, A., Huo, X., Kamel, M., Landrigan, P. J., Magalini, F., Diaz-Barriga, F., Neira, M., Omar, M., Pascale, A., Ruchirawat, M., Sly, L., ... Suk, W. A. (2016). E-waste and harm to vulnerable populations: A growing global problem. *Environmental Health Perspectives*, *124*(5), 550–555. https://doi.org/10.1289/ehp.1509699

Hischier, R., Reale, F., Castellani, V., & Sala, S. (2020). Environmental impacts of household appliances in Europe and scenarios for their impact reduction. *Journal of Cleaner Production*, *267*, 121952. https://doi.org/10.1016/j.jclepro.2020.121952

Höök, K., & Löwgren, J. (2021). Characterising interaction design by its ideals: Discipline in transition. *She Ji: The Journal of Design, Economics, and Innovation*, *7*(1), 24–40. http://doi.org/10.1016/j.sheji.2020.12.001

Ishii, H., Ratti, C., Piper, B., Wang, Y., Biderman, A., & Ben-Joseph, E. (2004). Bringing clay and sand into digital design: Continuous tangible user interfaces. *BT Technology Journal*, *22*(4), 287–299.

Ishii, H., & Ullmer, B. (1997). Tangible bits: Towards seamless interfaces between people, bits and atoms. In *Proceedings of the ACM SIGCHI conference on human factors in computing systems (CHI' 97)* (pp. 234–241). Association for Computing Machinery. https://doi.org/10.1145/258549.258715

Ismail, H., & Hanafiah, M. M. (2020). A review of sustainable e-waste generation and management: Present and future perspectives. *Journal of Environmental Management*, *264*, 110495. https://doi.org/10.1016/j.jenvman.2020.110495

ITU. (2024). *The digital sector's environmental dilemma*. International Telecommunication Union. Retrieved March, 2025, from https://www.itu.int/hub/2024/11/the-digital-sectors-environmental-dilemma/

Juma Nyeko, S., Mlay, V., Amerit, B., Abima, B., Among, J., Nyero, A. I., Odiya, J., & Ogen, C. (2023). The impact of electronic-electrical waste on human health and environment: A systematic literature review. *Journal of Engineering and Technological Research*, *15*(1), 1–16. https://doi.org/10.5897/JETR2021.0726

Karduri, R. K. R. (2023). Cobalt in battery production: Implications for the mining community. *International Journal of Advanced Research in Basic Engineering Sciences and Technology (IJARBEST)*, *10*(11), 1–7. https://doi.org/10.13140/RG.2.2.19114.80327

Kaunda, R. B. (2020). Potential environmental impacts of lithium mining. *Journal of Energy & Natural Resources Law*, *38*(3), 237–244. https://doi.org/10.1080/02646811.2020.1754596

Köhler, A. R. (2012). Material scarcity: A reason for responsibility in technology development and product design. *Science and Engineering Ethics*, *19*(3), 1165–1179. https://doi.org/10.1007/s11948-012-9401-8

Krishnan, R., & Gopan, G. (2024). A comprehensive review of lithium extraction: From historical perspectives to emerging technologies, storage, and environmental considerations. *Cleaner Engineering and Technology*, *20*, 100749. https://doi.org/10.1016/j.clet.2024.100749

Kulkarni, A., & Saraf, C. (2019, December). Learning from nature: Applications of biomimicry in technology. In 2019 *IEEE Pune Section International Conference (PuneCon)* (pp. 1–6). IEEE. https://doi.org/10.1109/PuneCon46936.2019.9105797

Liu, K., Tan, Q., Yu, J., & Wang, M. (2023). A global perspective on e-waste recycling. *Circular Economy*, *2*(1), 100028. https://doi.org/10.1016/j.cec.2023.100028

Lu, T., Chen, W.-Q., Ma, Y., Qian, Q., & Jia, J. (2023). Environmental impacts and improvement potentials for copper mining and mineral processing operations in China. *Journal of Environmental Management*, *342*, 118178. https://doi.org/10.1016/j.jenvman.2023.118178

MacMurdo, M. G., Mroz, M. M., Culver, D. A., Dweik, R. A., & Maier, L. A. (2020). Chronic Beryllium disease: Update on a moving target. *Chest*, *158*(6), 2458–2466. https://doi.org/10.1016/j.chest.2020.07.074

McCollough, J. (2009). Factors impacting the demand for repair services of household products: The disappearing repair trades and the throwaway society. *International Journal of Consumer Studies*, *33*(6), 619–626. https://doi.org/10.1111/j.1470-6431.2009.00793.x

Meyrowitz, J. (1986). *No sense of place: The impact of electronic media on social behavior.* Oxford University Press.

Mrozik, W., Rajaeifar, M. A., Heidrich, O., & Christensen, P. (2021). Environmental impacts, pollution sources and pathways of spent lithium-ion batteries. *Energy & Environmental Science*, *14*, 6099–6121. https://doi.org/10.1039/D1EE00691F

Mullai, B. D., & Sivasamy, R. (2017). Impact of vampire power and its reduction techniques – A review. In 2017 *International Conference on Intelligent Computing and Control Systems* (ICICCS) (pp. 404–405). Madurai, India. https://doi.org/10.1109/ICCONS.2017.8250752

Nagase, Y., & Uehara, T. (2024). The potential impact of the new 'Right to Repair' rules on electrical and electronic equipment waste: A case study of the UK. *Waste Management*, *182*, 175–185. https://doi.org/10.1016/j.wasman.2024.04.032

Parajuly, K., Kuehr, R., Awasthi, A. K., Fitzpatrick, C., Lepawsky, J., Smith, E., & Zeng, X. (2019). *Future e-waste scenarios.* Retrieved March, 2025, from http://collections.unu.edu/eserv/UNU:7440/FUTURE_E-WASTE_SCENARIOS_UNU_190829_low_screen.pdf

Raabe, D. (2023). The materials science behind sustainable metals and alloys. *Chemical Reviews*, *123*(5), 2436–2608. https://doi.org/10.1021/acs.chemrev.2c00799

Ray, M., & Jat, K. R. (2010). Effect of electronic media on children. *Indian Pediatrics*, *47*(7), 561–568. https://doi.org/10.1007/s13312-010-0128-9

Rene, E. R., Sethurajan, M., Kumar Ponnusamy, V., Kumar, G., Bao Dung, T. N., Brindhadevi, K., & Pugazhendhi, A. (2021). Electronic waste generation, recycling and resource recovery: Technological perspectives and trends. *Journal of Hazardous Materials*, *416*, 125664. https://doi.org/10.1016/j.jhazmat.2021.125664

Resch, B., Britter, R., & Ratti, C. (2012). Live urbanism – Towards SENSEable cities and beyond. In S. T. Rassia & P. M. Pardalos (Eds.), *Sustainable environmental design in architecture* (pp. 175–184). Springer. https://doi.org/10.1007/978-1-4419-0745-5_10

Schön-Blume, N., Dolega, P., Buchert, M., Bodenmüller, B., & Meyer, M. (2021). *The impact of gold: Sustainability aspects in the gold supply chains and Switzerland's role as a gold hub*. Oeko-Institut e.V. and WWF Switzerland. Retrieved March, 2025, from https://www.wwf.ch/sites/default/files/doc-2021-11/2021-11-WWF-The-Impact-of-Gold.pdf

Selvakumar, S., Adithe, S., Isaac, J. S., Pradhan, R., Venkatesh, V., & Sampath, B. (2023). A study of the Printed Circuit Board (PCB) E-waste recycling process. In A. M. Rawani, M. K. Sahu, & S. S. Chakarabarti (Eds.), *Sustainable approaches and strategies for E-waste management and utilization* (pp. 26–44). IGI Global. https://doi.org/10.4018/978-1-6684-7573-7.ch009

Shakoor, N., Adeel, M., Ahmad, M. A., Zain, M., Waheed, U., Javaid, R. A., Haider, F. U., Azeem, I., Zhou, P., Li, Y., Jilani, G., Xu, M., Rinklebe, J., & Rui, Y. (2023). Reimagining safe lithium applications in the living environment and its impacts on human, animal, and plant systems. *Environmental Science and Ecotechnology*, *15*, 100252. https://doi.org/10.1016/j.ese.2023.100252

Shirvani, G., & Ghasemshirazi, S. (2024). Towards sustainable IoT: Challenges, solutions, and future directions for device longevity. *ArXiv*. https://doi.org/10.48550/arXiv.2405.16421

Siragusa, C., & Tumino, A. (2021). E-grocery: Comparing the environmental impacts of the online and offline purchasing processes. *International Journal of Logistics Research and Applications*, *25*(8), 1164–1190. https://doi.org/10.1080/13675567.2021.1892041

Sonter, L. J., Ali, S. H., & Watson, J. E. M. (2018). Mining and biodiversity: Key issues and research needs in conservation science. *Proceedings of the Royal Society B: Biological Sciences*, *285*(1892), 20181926. https://doi.org/10.1098/rspb.2018.1926

Subramanian K. & Yung W. K. C. (2016). Review of life cycle assessment on consumer electronic products: Developments and the way ahead. *Critical Reviews in Environmental Science and Technology*, *46*(18), 1441–1497. https://doi.org/10.1080/10643389.2016.1245550

Tran, B. X., Ha, G. H., Vu, G. T., Hoang, C. L., Nguyen, S. H., Nguyen, C. T., ... Ho, R. C. M. (2020). How have excessive electronic devices and Internet uses been concerned? Implications for global research agenda from a bibliometric analysis. *Journal of Behavioural Addiction*, 9(2), 469–482. https://doi.org/10.1556/2006.2020.00031

Van den Berge, R., Magnier, L., & Mugge, R. (2023). Sparking the repair "can-co" attitude: Enhancing users' willingness to repair through design support in fault diagnostics. *International Journal of Design*, *17*(3), 25–39. https://doi.org/10.57698/v17i3.02

Verma, A. (2020). E-wastes and their impact on environment and public health. *International Journal of Applied Research*, *6*(9), 164–168. Retrieved March, 2025, from https://ssrn.com/abstract=3699964

Weerasundara, L., Mahatantila, K., & Vithanage, M. (2020). E-waste as a challenge for public and ecosystem health. In M. N. V. Prasad, M. Vithanage, & A. Borthakur (Eds.), *Handbook of electronic waste management* (pp. 101–117). Butterworth-Heinemann.

CHAPTER 2

BIODERIVED MATERIALS FOR ELECTRONIC AND INTERACTIVE DEVICES

DOI: 10.4324/9781003366799-3

A BIODERIVED PERSPECTIVE FOR ELECTRONICS AND INTERACTIVE DEVICES

It is well known that the materials traditionally used in digital technologies have a significant environmental impact throughout their lifecycle. This awareness has catalyzed a growing body of research focused on exploring alternative, more sustainable materials – particularly biobased, biofabricated, and living materials – for interactive and electronic devices. The term bioderived is used here as an overarching category that encompasses these three types of materials. Over the past few decades, bioderived materials have experienced a resurgence of interest in fields as diverse as fashion and packaging and the more unexpected realm of electronics (Mühl & Beyer, 2014). Their study requires interdisciplinary engagement between materials science, chemistry, biotechnology and design. It exploits the unique properties of biological building blocks – such as proteins, polysaccharides and minerals – that underpin many natural systems and structures.

The appeal of biobased and living materials lies in their perceived sustainability. Derived from renewable biological sources, these materials are generally abundant and biodegradable, provided they are not contaminated by synthetic additives. However, the field is still in its infancy, with limited scientific data on the environmental impact of biofabrication processes and the challenges of scalability. Crucially, material substitution alone is unlikely to advance sustainability transitions significantly. Instead, design must incorporate targeted strategies – such as durability-focused approaches and the promotion of behavioral change – to reduce the overall footprint of electronic products. In this context, the expanding field of biodesign explores the integration of living organisms as functional design elements (Myers, 2012). While many case studies remain speculative or experimental, they illustrate emerging sustainability paradigms in which the regenerative processes inherent in living systems are harnessed in material and interactive design. Increasing efforts are being made to preserve the vitality of these organisms to the point where "*livingness*" is being recognized as a "material quality" (Karana et al., 2020). Defining the degrees and characteristics of livingness is critical for articulating design practice's technical and ethical dimensions.

Preserving organisms' vitality helps harness their inherent capabilities as functional components. It is imperative to establish evident characteristics and definitions that provide a framework for delineating the varying degrees of livingness exhibited by the organisms involved. This aspect has profound implications in the field of interactive technologies. While organic but inert materials can be perceived as material substitutes (with active responses to the environment based on their physical and chemical composition), living organisms can serve as biosensors or cleaning technologies, challenging our conventional understanding of "interactive design". When living organisms are integrated into technologies as functional components, the field enters the realm of biological human–computer interaction (bioHCI), which includes a comprehensive framework that uses biological elements as materials for human–computer interfaces (Pataranutaporn, 2018). With a particular focus on living interfaces, Merritt et al. (2020) have conducted a comprehensive analysis of living media interfaces (LMIs), which are digital systems whose interactive capability depends on the incorporation of living organisms as user interfaces. Their study provides a panoramic view of the current design landscape, highlighting open research questions from four perspectives – biological, ethical, artistic and human–computer interaction (HCI). It highlights user engagement through shared vitality, temporal patterns guided by biology, cultural significance and the imperative of multidisciplinary expertise, advocating for robust collaboration between researchers, designers and artists (Merritt et al., 2020).

Scholars agree that this expanding range of biological and living material possibilities in interaction design should not be interpreted as limiting the interactive experience to prioritize sustainability.

Instead, it should be considered as an opportunity to redefine new models of interaction, as articulated by Parkes and Dickie (2013). They propose that all matter, both living and non-living, should be viewed as a hardware system based on its ability to sense, derive inputs, interpret them and generate feedback. This way of thinking is presented as a universal approach for incorporating biological processes into interaction scenarios (Parkes & Dickie, 2013). This approach aligns harmoniously with established language and contemporary knowledge in interactive and electronic devices and is proving to be crucial in the field of interactive digital media.

EMERGING TERMINOLOGIES AND CLASSIFICATIONS IN BIODESIGN

The rapid growth of interest and research in biodesign has left little time to establish a consolidated and widely used glossary. As a result, terminology relating to materials derived from the generative activity of living organisms remains vague and inconsistent. Such materials have been variously referred to as "biomaterials" – a term that originated in the medical field of tissue engineering; as "growing materials" (Camere & Karana, 2017), referring to the morphogenetic capabilities of organisms used in biodesign; or more broadly as materials "made of, with or from living organisms" (Ginsberg & Chieza, 2018). In 2020, the report *Understanding Bio* (Lee et al., 2020) further clarified the glossary of biobased materials, building on the earlier work of Mironov and Groll (Mironov et al., 2009; Groll et al., 2016). In many disciplines, the term "biobased" refers to where the material comes from, suggesting a natural origin. The *Understanding Bio* report proposes the broad definition of "biobased" for all those materials that are "derived in whole or in part from biomass". Under this broad umbrella, the report provides another, more specific category, referring to "biofabricated materials" as those produced by living cells or microorganisms. The report describes "biofabricated ingredients" as building blocks made by living cells or microorganisms that "require further mechanical and chemical processing" into materials and structures. In addition, macroscale structures grown directly from living microorganisms are defined as "bioassembled materials". Most biofabricated materials have been produced by microorganisms and stabilized to become inert for the final production and use phase. However, the possibility of keeping the organisms alive is increasingly being explored, leading to another definition, "living materials" (Karana, 2020), which describes all those substances produced by cells and organisms that are still living and/or kept alive for design purposes. As the definition of life itself is currently controversial, the distinction between living and non-living remains a non-trivial concept. Rasmussen and colleagues point out that "*there is general agreement that a localised molecular assemblage should be considered alive if it is capable of continuously regenerating, replicating and evolving*" (Rasmussen et al., 2004). To clarify the potential that different stages of an organism's vitality can bring to the design process, we can refer to the taxonomic scale developed by Pollini (Pollini, 2021, 2023), which considers how the inert/living state influences design possibilities and outcomes, to clarify the aspects related to the life stage of the organism, which in biodesign can give life to materials with very different properties and abilities to interact.

This taxonomic scale, modeled on a broad analysis of biodesign case studies, highlights two poles with different characteristics and meanings. This duality of the taxonomic scale corresponds to a binary aspect of life embedded in design: freedom of expression (right pole of the scale) and confinement (either for the organism's death or its lack of freedom of expression). Such an on/off expressivity of the organism is a determinant of different design possibilities, values and designer and user experiences, as well as philosophical and ethical concepts that are radically divergent. Depending on the expressiveness of the organism, the taxonomy outlines these two poles through five key aspects of the project, contrasting with each other based on the organism's living conditions, as

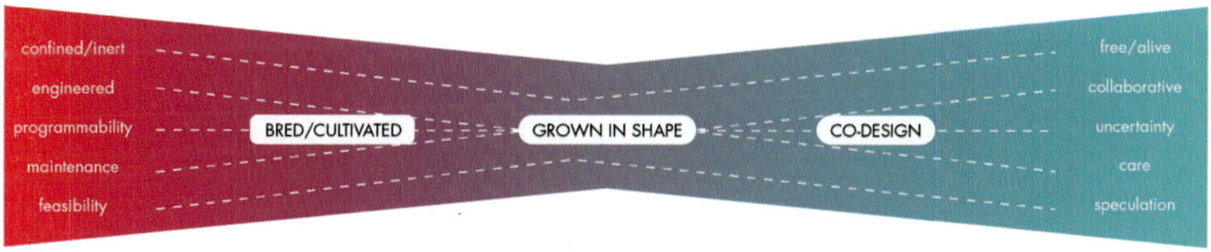

Figure 2.1 Healing materialities taxonomic scale (Pollini, 2023).
Credits: Barbara Pollini.

shown in Figure 2.1. The left side of the scale describes materials and organisms of biological origin that are i) produced by living organisms but suddenly stabilized by the death of the organisms, becoming inert materials; or ii) limited living materials, namely those living entities that, despite their aliveness, are designed for highly predictable outcomes. In this case, the interaction of the material is based on physical/chemical reactions derived from a material now deprived of its vital expressiveness. This condition determines the possibility of engineering materials and living organisms with predictable outcomes, making it easier to imagine the feasibility of their short-term application. On the right side of the scale are living organisms with more freedom of expression, materials that change, mutate and evolve during use, expressing themselves according to cognition and the stimuli of the perceived environment. Here, the design outcomes are challenging to predict as they result from a co-design process performed with the organisms. Since these organisms are alive, they may need maintenance during the use phase, which could quickly turn into "care and affection", leading to a relational attitude similar to the one we have with houseplants or pets. At the moment, this possibility is being explored at a somewhat speculative level, given the impossibility of obtaining predictable and specific results; however, nothing should prevent us from thinking that the integration of unpredictability will become a new valid parameter in future productions, bringing this aspect ever closer to design feasibility and multispecies environments (Pollini, 2023).

BIODERIVED BUILDING BLOCK FOR INTERACTIVE APPLICATIONS

Life on Earth fundamentally depends on materials to carry out many functions. Thanks to advances in diagnostics and biofabrication technologies, we are rapidly expanding our understanding of the ability of living organisms to manipulate and self-generate matter. This includes novel and innovative substances and familiar materials such as wood or leather, which can now be redefined as biobased and biogenerated, and more broadly as bioderived. As discussed in Chapter 1, the technology field has thus far been dominated by human-made and non-renewable materials, which are challenging to manage due to their environmental impact and difficulties in recovery at the end of their lifecycle. However, conventional biobased materials can also be reimagined through innovative applications to meet the demands of modern technology.

New bioderived materials, developed from nature's building blocks, can serve as sustainable alternatives to conventional fossil-based materials. In addition, research into these materials provides an opportunity to reduce dependence on rare materials, the extraction of which is often associated with significant environmental degradation and social impacts, including labor exploitation and community displacement. Nature's building blocks, responsible for creating matter and structure in all life forms, can be broadly categorized into minerals, proteins and sugars. Living organisms depend on their capacity to generate and regenerate these limited yet essential resources to construct shelters, form new body parts, fashion hunting weapons and more. The study conducted by Eder, Shahrouz and Fratzl (Eder et al., 2018) illustrates this array of material possibilities in a graph (as shown in Figure 2.2), demonstrating how living organisms derive complex shapes and structures for a wide range of applications from a few simple, abundant and renewable sources.

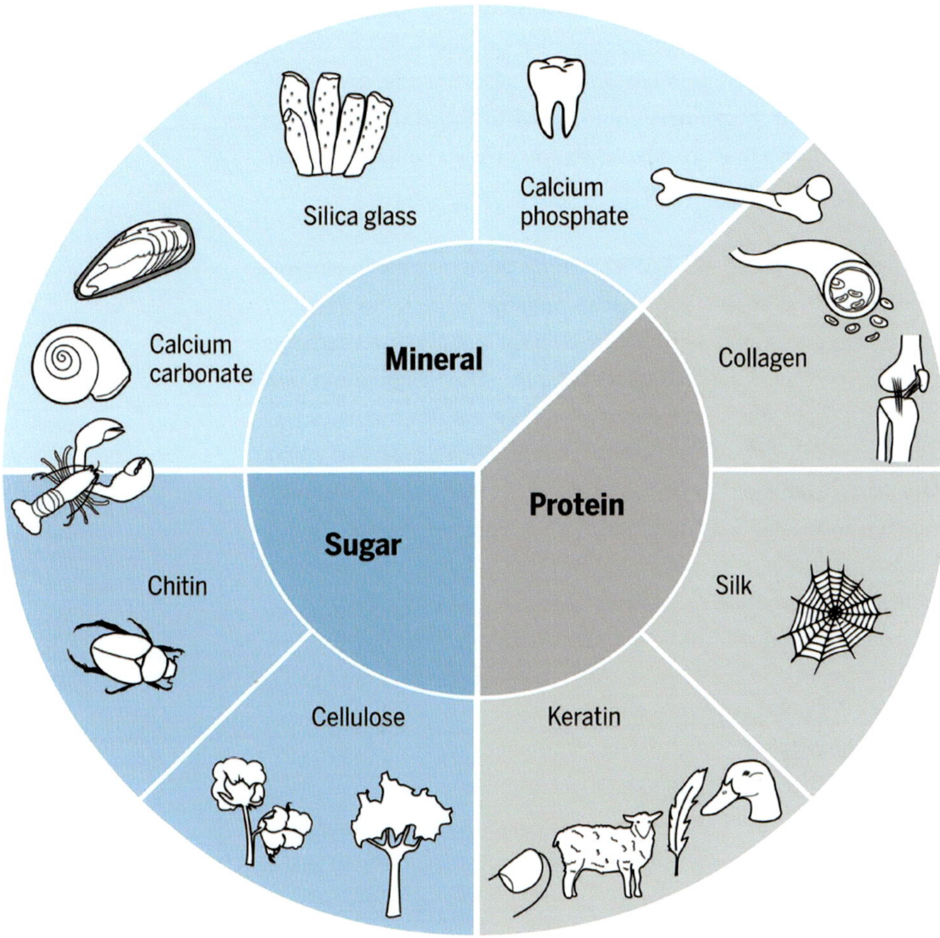

Figure 2.2 Biological materials are built with limited building blocks based on polysaccharides, proteins and minerals. Credits: (Eder et al., 2018).

In this comprehensive analysis, we explore the range of biobased materials, structures and derivatives revolutionizing the interactive and electronic device industry and the digital technology sector, focusing on their transformative impact on interaction design. Our discussion encompasses these natural components' vital role in forging a new generation of sustainable and biobased interactive systems, illustrating their potential to foster innovation and cultivate more than human-centered products. By exploring these groundbreaking materials and their applications, we aim to demonstrate their significant contribution to the development of interactive artifacts that are innovative and aligned with sustainable practices.

In the upcoming sections, we will explore a wide range of biological materials central to several fields, ranging from biotechnology and regenerative medicine to producing sustainable materials for the electronics and digital sectors. Our exploration will cover the properties, applications and potential for innovation of polysaccharides, including cellulose, bacterial cellulose and chitin. We will then examine protein-based materials alongside DNA, such as keratin and collagen. In addition, we will highlight the importance of active pigments – substances that can change their color or properties in response to external stimuli – as a significant step forward in creating smart and adaptive materials.

Polysaccharides

Polysaccharides are common biopolymers widely distributed in nature and they consist of simple sugars (monosaccharides) linked by glycosidic bonds (Udayan et al., 2017). These biomolecules play a dual role in nature as energy sources (e.g. starch and inulin) and structural building blocks (e.g. cellulose and chitin). They also have a wide range of applications in the human-made environment, including materials, food and pharmacy/medicine. The study of polysaccharides has led to a deeper understanding of their complexity in nature and significant innovations in various fields, particularly in developing new biobased materials (Persin et al., 2011).

Polysaccharides can exhibit a wide range of mechanical properties, from rubbery to glassy. However, they can be susceptible to degradation at elevated temperatures and often exhibit limited thermoplastic behavior. In addition, hydrated polysaccharides are susceptible to microbial attack and biodegradation, factors that must be considered in the product's lifecycle (Persin et al., 2011). Polysaccharide-based materials have attracted considerable attention in various sectors, from food packaging to biomedical applications. Their inherent antimicrobial properties have proven invaluable in creating scaffolds, wound dressings, drug delivery systems and antimicrobial coatings (Souza et al., 2021). These materials also provide solutions for microbial growth prevention and food safety in packaging (Altaf et al., 2022). Polysaccharide-based materials have even made inroads into the field of electronics, serving as dielectric materials, transistors and semiconductors (Sun et al., 2017). Ongoing research in the field aims to address these limitations and develop effective solutions, although challenges such as water solubility and limited thermal stability remain.

In the following subsections, we will explore abundant and inexpensive sources of polysaccharides, including cellulose, chitin, chitosan, and alginates derived from algae.

Cellulose

Cellulose, the most abundant polysaccharide in nature, is found in many organisms, including plants, marine animals, algae, fungi and bacteria. It is widely available, inexpensive, renewable, lightweight, biocompatible and biodegradable, among other remarkable properties. These properties have led to extensive research into the development of new biobased technologies and materials.

Particular attention has been paid to two prominent forms of cellulose: plant cellulose and bacterial cellulose. Wood, which is mainly composed of cellulose, has already been identified as a sustainable alternative material in the field of electronics. Malik and colleagues have investigated wood for flexible electronics and highlighted its diverse properties, including chemical, mechanical, optical, thermal, thermomechanical and surface properties, which have applications in electronics. Furthermore, developing a conductive ink based on lignin using translucent wood as a substrate could increase the use of plant-based materials in electronic devices (Malik et al., 2023). Besides wood, plant cellulose can be extracted from cotton, bamboo and rice husks. As a significant component of plant cell walls, it is a naturally stable and non-toxic material (Naomi et al., 2020). Research projects on plant-based cellulose have demonstrated its extreme versatility and adaptability for design, as evidenced by the Chemarts cookbook[1] of student recipes developed by the Design School of Aalto University. Furthermore, researchers at the University of British Columbia[2] highlight the circular potential of plant cellulose in the bioeconomy by exploring the use of fibrillated cellulose from crops and other biomass wastes to create materials ranging from composites and macrofibers to thin films, porous membranes and gels (Li et al., 2021).

In terms of interactive smartness, cellulose has a sensitivity to water that enables intelligent material responses to moisture. The pinecone effect is a well-known biomimetic example derived from the ability of pinecones to open and close in response to changes in humidity. Natural and artificial hygromorphs are materials or objects that change shape in response to environmental moisture (Reyssat & Mahadevan, 2009). Several principles of plant movement can be applied to hydraulically actuated 3D-printed hygroscopic structures (Poppinga et al., 2020). For example, experiments on programmable wood take inspiration from its anisotropic behavior, resulting from irregular grain patterns, to create 3D-printed customisable grain patterns with predictable wood bending responses to moisture and heat. This bioinspired material transformation mechanism can also be applied to other materials, such as carbon fibers or textiles, as demonstrated by the MIT Self-Assembly[3] experiments on programmable materials (Papadopoulou et al., 2017).

As living organisms, plants exhibit material transformation in inert parts (e.g., reactive seed scaffolds), through morphogenesis (the growth of new body parts) and in response to external stimuli (e.g., dormant movements of flowers in response to light intensity or contraction reactions to tactile stimuli). Plant movement can be seen as an exciting example of living and interactive material behavior: seed dispersal units, climbing plants and carnivorous plants express shape change as a predominant function "by using cellulose microfibrils in specific orientations to anisotropically constrain the swelling/shrinking of their organic matrices in response to external stimuli" (Erb et al., 2013).

Bacterial Cellulose

One of the most promising resources for the future of biobased interactive devices is bacterial cellulose (BC). BC combines high crystallinity, exceptional water-holding capacity, and biocompatibility, making it a sustainable platform for value-added applications from packaging to biomedical devices (Gallegos et al., 2016). The introduction of BC, produced by *Acetobacter xylinum*, dates back to 1886 when it was first described by the scientist Brown. Since then, numerous bacterial strains have been identified as cellulose producers.

Designers have taken a keen interest in these materials, mainly inspired by the results of Suzanne Lee's biocouture research at Central Saint Martin's College of Art and Design in the early 2000s. Lee used a culture of kombucha-fermenting bacteria and yeasts to create a protective layer of cellulose that naturally self-assembles into a non-woven textile (Lee, 2005). This biofilm, also known as SCOBY (symbiotic culture of bacteria and yeast), is easy to produce even in non-high-tech environments, making it a popular choice for DIY-materials experimentation (Rognoli et al., 2015; Rognoli & Ayala-Garcia, 2021).

Researchers at the Design F(r)iction Lab of the Free University of Bozen-Bolzano have explored the production (Cohen et al., 2023) and potential uses of BC as a circular edible and non-edible material (Cohen et al., 2023; Sicher, 2017). The interdisciplinary investigation included local industrial byproducts as a source of nourishment for the microbial culture to envision regional circular production scenarios (Cohen et al., 2019). Packaging has been the primary focus for the development of applications; however, favorable preliminary evaluations by food microbiologists have set the ground for developing food concepts.[4] The InnoCell Bioreactor was designed to enable the production of enough BC for the exploration of material prototypes – ranging from powder and granulate to sheets and solid foam – and the application prototypes.

Similarly, the Politecnico di Milano has undertaken several research projects on materials experimentation and biotinkering with BC derived from kombucha fermentation. The DeForma research project[5] investigates the design potential of BC by experimenting with the growing phase (e.g., liquid c composition, testing circular carbon sources, etc.), hybridizing BC with other materials and components and exploring potential applications in wearable fashion tech and healthcare. The research group has created a gallery of samples, exploring established and novel DIY techniques (Pollini, 2023), analyzing the expressive-sensorial dimension of BC-developed samples (Papile et al., 2023) and supporting other researchers and graduate students interested in experimenting with this organism. Among these researchers is Samuele Sergio Fontana, who has studied five production techniques and different cultivation practices (Figure 2.3) to "facilitate the understanding and dissemination of information related to the use of BC in design" (Fontana, 2023). Ziqian Yu is a PhD candidate at Polimi dedicated to researching biofabricated materials for use in healthcare and promoting overall well-being. Specifically, she focuses on developing bacterial cellulose with healthcare functions (Yu, 2023; Yu et al., in press). Inspired by the rotating disk bioreactor described in the Bungay et al. (1999) patent and building on the

Figure 2.3 Five production techniques for bacterial cellulose (BC), developed by Samuele Fontana through experimental research during his Master's degree in Design and Engineering at the Politecnico di Milano (2023).
Credits: Samuele Fontana.

InnoCell project developed at the Design F(r)iction Lab, she has designed and implemented a bioreactor for cultivating bacterial cellulose. She is currently exploring alternative solutions to improve and diversify bacterial cellulose (BC) production systems.

From a design perspective, BC is becoming an increasingly important research topic in the field of human–computer interaction (HCI). Researchers and designers are exploring using SCOBY as a material for bio-HCI. Bell and colleagues propose SCOBY for material exploration activities to create interactive wearables and applications (see Figure 2.4 and Figure 2.5). They emphasize one of the most distinctive qualities of biological interfaces, which is slowness. The authors discuss the different time scales involved in the SCOBY's life cycle and encourage designers to make connections between the concepts of slow design, bio-HCI and sustainable HCI (Bell et al., 2023). This research demonstrates the potential of BC and SCOBY in HCI, particularly in the context of sustainable and bioinspired design.

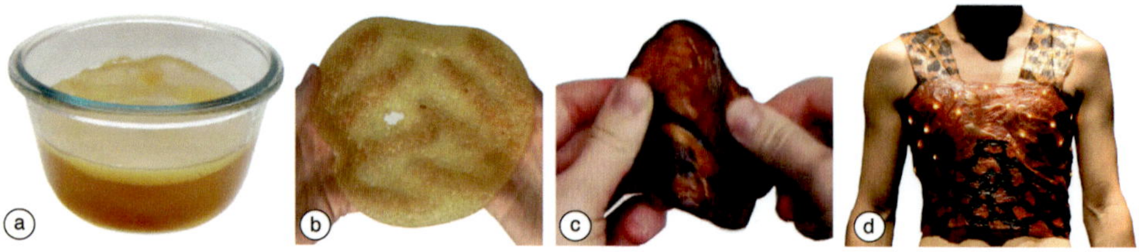

Figure 2.4 A kombucha SCOBY (a symbiotic culture of bacteria and yeast) is a sustainable cellulosic biofilm which is cultured in kombucha tea and can subsequently be harvested and dried. When dried, the SCOBY mimics the properties of traditional animal and synthetic leathers. This studio is designed to explore the customisation of dry SCOBY 'leather' through colour, shape and electronics, with the aim of developing wearables and other interactive applications for human–computer interaction (HCI).

Credits: Bell et al., 2023.

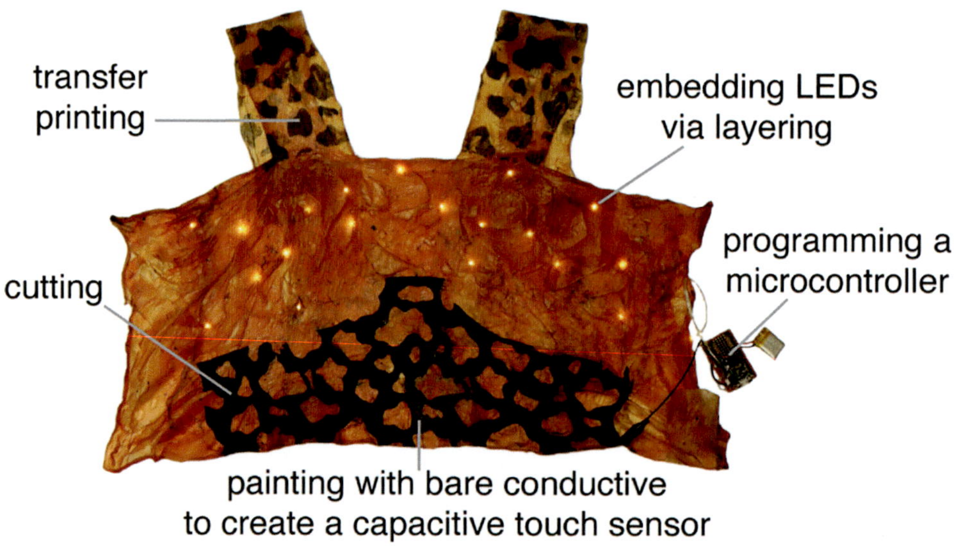

Figure 2.5 The image showcases an open exploration of SCOBY, in which participants learn various fabrication techniques, including cutting, layering, transfer printing, embedding LEDs, and enhancing electrical conductivity. This phase precedes a structured activity with SCOBY, guiding participants to explore the creation of an interactive SCOBY-based wearable or application.

Credits: Bell et al., 2023.

In a more recent study (Bell et al., 2024), the same researchers explored how active engagement with the lifecycles of non-human organisms can foster an understanding of multispecies interactions. To do this, they created a biodigital calendar highlighting the growth and well-being of the Kombucha SCOBY, a symbiotic culture of bacteria and yeast that thrives in a tea medium. This non-invasive biodigital calendar encases the Kombucha SCOBY and uses sensors to monitor and sounds to encourage its growth. By regularly observing and listening to the calendar of the Kombucha SCOBY, people can become more aware of and in tune with the gradual development of the Kombucha SCOBY.

The appeal of bacterial cellulose to designers and researchers lies in its unique material properties, primarily related to its structure. Bacterial cellulose consists of a three-dimensional open network of intertwined cellulose nano and microfibrils. This intricate structure offers several advantages and potential applications.

In addition, researchers have identified numerous wild and recombinant strains of bacteria capable of producing cellulose. These strains can use various carbon sources as substrates, including food and agricultural wastes. This versatility in carbon sources allows for flexibility in the production process, making bacterial cellulose an environmentally friendly and sustainable material option for various applications.

Bacterial cellulose is produced by gram-negative[6] bacteria such as *Komagataeibacter (Gluconacetobacter), Agrobacterium, Achromobacter, Enterobacter, Rhizobium, Pseudomonas, Salmonella, Azotobacter and Alcaligenes, and gram-positive bacteria* such as *Sarcina ventriculi* and *Rhodococcus*. Among these, the best-known cellulose-producing bacterium is the acetic acid bacterium *Komagataeibacter xylinus*, also known as *Gluconacetobacter xylinus, Acetobacter xylinum, A. aceti ssp. xylinum,* or simply *A. xylinus* (Revin, Liyas'kina, Sapunova et al., 2020).

In addition to the commonly used strains, bacteria and yeasts can be genetically engineered to develop various mutant strains that improve BC's structural properties and productivity (Singhania et al., 2021).

In contrast to plant cellulose, bacterial cellulose is purer and is free of substances such as hemicellulose, lignin or pectin. In addition to its purity, bacterial cellulose exhibits excellent mechanical properties, high crystallinity and remarkable water-holding capacity (Jang et al., 2017). These properties make it suitable for applications in tissue engineering, drug delivery, cosmetics, food and polymer nanocomposites. Incorporating inorganic or organic fillers further expands the utility of BC, for example, in food packaging, thermo-responsive materials, biomedical device fabrication and more (Blanco et al., 2018).

BC's strength and flexibility make it a promising replacement for conventional petrochemical-based polymers in flexible electronics, potentially reducing non-biodegradable polymer waste. BC-based electronic applications often involve composites of bacterial nanocellulose and electrically conductive materials such as conductive polymers, metal and carbon nanoparticles. Common conductive fillers encapsulated in the bacterial cellulose matrix to enhance electrical conductivity include polyaniline (PANI), polyacetylene (PAc) and polypyrrole (PPy), along with materials such as metal nanoparticles, metal oxides, carbon nanotubes, graphene and carbon nanofibers (Hosseini et al., 2021). The hydrophilic nature of BC makes it compatible with hydrophilic polymeric materials or nanofillers (Poddar & Dikshit, 2021). BC composites created by encapsulating conductive fillers have potential applications in manufacturing electrical and electronic devices, biosensors, fuel cells, electromagnetic interference (EMI) shielding, supercapacitors, electronic cards and actuators. Another promising application of BCs in interactive devices is as power generators. Kim et al. (2017) developed a BC-based biotriboelectric nanogenerator that can convert mechanical energy from the environment into electricity, which can power small devices such as sensors or charge consumer electronics (Kim et al., 2017). This new BC application could be a sustainable energy source for remote-sensing electronics, autonomous devices or portable self-power generators.

BC-based devices can also have tunable properties. Onggo and Mulyani (2017) induced the growth of microcrystals of a one-dimensional spin-crossover compound within a BC matrix, resulting in reversible thermomagnetic and thermochromic stability in the composite films. These properties, including a transition from diamagnetic to paramagnetic behavior and a color change from purple to pale yellow, exhibited a well-defined hysteresis in the temperature range of about 345–380°K. Furthermore, these properties were influenced by the history of the sample, demonstrating the composite memory effect (Onggo et al., 2017).

BC, in the form of hydrogels, 3D aerogels and nanofiber dispersions, can be used as a substrate, reinforcing agent and active material to create flexible physical sensors. These sensors have applications in human motion monitoring, human–machine interfaces, intelligent robotics and personalized medicine. In addition, the hygroscopic nature of BCs can be used to accumulate and release aromas and essential oils for contamination control and as active packaging for food preservation (Indrati, 2017; Junka et al., 2019); this might pave the way also for such BC applications in multi-sensory devices. Another fascinating attribute of BC is its edibility, which makes it applicable in the emerging field of edible electronics (Ferrarese et al., 2023; Prilepskii et al., 2023; Azeredo et al., 2019).

All properties highlighted in this section contribute to BC's attractiveness as a promising biomaterial for numerous applications. These qualities and characteristics broaden the horizons for designing devices with biobased, low-cost, renewable and biodegradable components, opening new possibilities in various industries.

Chitin

Chitin is the second most abundant polysaccharide after cellulose. It is the primary structural material in the exoskeletons of crustaceans and arthropods and is also found in certain fungi and algae. Like cellulose, chitin is biocompatible, easy to source and process and has antimicrobial properties. These properties have led to the widespread use of chitin in the technology sector as a flexible substrate to replace petroleum-derived materials. For example, it has been tested as an active component in combination with graphene, metal oxides and ionic liquids. These combinations have resulted in humidity, temperature and gas sensors, lightweight and environmentally friendly energy storage devices, electronic components and electrical circuits (Pradhan et al., 2020).

One of the most promising applications of chitin in wearable electronics and soft robotics is the development of chitin-based hydrogels, known as CTAs. Zhang et al. (2023) have detailed the design and characterization of a class of non-covalently cross-linked multifunctional hydrogels with excellent overall performance. These hydrogels result from cross-linking carboxyethyl chitin (CECT) and acrylamide (AAm). These materials are easy to prepare, sustainable and can incorporate multiple functions. In this research, wearable strain sensors based on CTA hydrogels have been developed to monitor body movements. These flexible sensors exhibit stretchability, high sensitivity, low detection limits, fast response times, and exceptional reliability and durability. In addition, the same research has led to the creation of soft and adaptable bioelectronic devices, including a human–machine interactive system, an electronic skin for non-planar pressure sensing, and self-charging energy harvesting devices (Zhang et al., 2023).

The deacetylated form of chitin, known as chitosan, is also a promising material for interactive devices. Chitosan has unique properties, such as much higher solubility than conventional solvents. However, some key challenges remain to overcome, including chitosan's high water vapor permeability and modest mechanical properties. Chitosan, similarly to bacterial cellulose, can be used in the fabrication of triboelectric nano-generating devices (TENGs). A study by Kim et al. developed a chitosan-diatom TENG for biofriendly and skin-attachable wearable devices to support a sustainable Internet of Things (Kim et al., 2017). Furthermore, using chitosan in electronic skin applications may herald a creative revolution in the field of wearable devices and tissue engineering (Lin et al., 2018).

Protein-Based Materials

Over the past few decades, scientists have been working to expand the applications and versatility of natural proteins. Natural, protein-based materials such as silk, collagen, keratin and elastin act as structural components in biological systems, offering a combination of strength, resilience and flexibility (Harrington & Fratzl, 2021). These protein-based materials, readily available and well suited to the circular economy, have proven to be biocompatible and biodegradable.

As a result, they are rapidly transforming various sectors, including medicine, electronics, short-life products and fashion. Some of the most commonly used proteins in biomaterials include silk, elastin, resilin, collagen and keratin, all characterized by their remarkable mechanical strength and elasticity (Abascal & Regan, 2018).

The study of natural, protein-based materials emphasises the close relationship between the structure and function of biological matter, a topic also covered by Fratzl (2021) in his lectures as part of the Matters of Activity cluster. This perspective strengthens the notion that biological systems could inform the development of programmable, adaptive and sustainable materials for future design and engineering applications.

Silk

Silk, a protein-based material, garners significant attention due to its remarkable versatility. In the natural world, silk is produced by spiders, certain wild moths and bivalves. However, for nearly 5,000 years, humans have domesticated the silkworm *Bombyx mori*, a member of the moth family *Bombycidae,* to cultivate silk. Silk comprises two primary components: 80% fibroin (a protein) and 20% sericin (a rubbery substance).

Widely known for its exquisite fabrics, silk and its core components have become subjects of experimentation for novel material

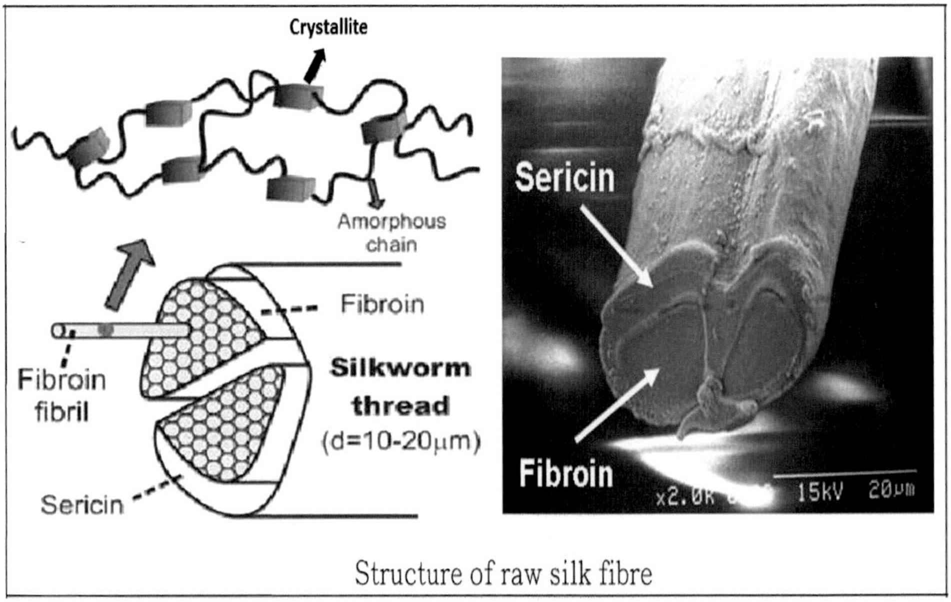

Figure 2.6 Schematic representation of the native silk fiber hierarchical structure. Credits: (Guidetti et al., 2022).

processing and applications. Silk and fibroin have attracted the interest of materials scientists due to their versatility and broad applicability in fields such as biomedical devices, optics, food packaging, design, and architecture (Guidetti et al., 2022). Among several structural proteins under investigation, fibroin is particularly notable for its potential to create biocompatible functional materials (see Figure 2.6). It can be processed at room temperature, and its programmable biodegradability, edibility, and biocompatibility make it a promising sustainable material. As a result, it is widely used in organic electronics and the medical sector, particularly for implantation in the human body without triggering an immune response and subsequent spontaneous resorption. The tunability of fibroin is due to the antiparallel β-sheets (often referred to as crystallites) formed by the glycine and alanine-rich repeat regions of the protein (Abascal & Regan, 2018). Silk offers a wide range of material formats, with results ranging from plastic-like materials to optical fibers, ceramic-like blocks, printable functional inks, 3D-printed geometries, porous materials, molded objects and nanostructured materials, to name a few (Omenetto, 2017).

Compared to other biopolymers, fibroin is an exceptional material for high-tech applications due to its robust mechanical properties, easy tunability of material properties by controlling the water content during processing, unique optical and electronic properties and programmable degradation lifetime. Fibroin has been extensively experimented with in photonic devices (Omenetto & Kaplan, 2008), organic electronics (Tao et al., 2014), flexible electronic devices (Zhu et al., 2016), bioactive inks and reactive surfaces (Matzeu et al., 2020) and silk coatings for active packaging (Marelli et al., 2016), to name a few. Furthermore, it has been described as a "living material" due to its dynamic biochemical behavior (Tao et al., 2014). As a structural building block of biological origin, fibroin is ideal for interfacing with organic agents. For example, silk films can preserve the function of biological dopants such as enzymes for over a year at room temperature (Lu et al., 2010). In addition, silk hydrogel structures have been tested to preserve the vital functionality of marine microalgae; fibroin structures supported cell proliferation for one month and maintained consistent photosynthetic activity for over 90 days (Zhao et al., 2019). This research shows that silk has the potential for a wide range of applications where biological vitality and functionality can be maintained.

Keratin

Keratin is a structural protein commonly found in nature in materials such as nails, hair, feathers, hooves, horns and claws. Its slow biodegradability often leads to its classification as a harmful waste. However, its abundance and unique physicochemical properties make it increasingly attractive for sustainable product applications (Pradhan et al., 2020).

Keratin has been used as a substrate and an active material for electrical transduction in moisture-sensing applications. In addition, human hair has been used in resistive memory devices (Lin et al., 2019). This material can also be exploited for its insolubility in water during processing and later treated to create transient devices (Kang et al., 2020).

Keratin is readily available as a waste stream from the textile (wool) and food (feathers) industries. Some Italian researchers at the IIT of Genoa (Cataldi et al., 2019) extracted keratin from waste wool and combined it with graphene to make resistive planar capacitors and inductors for analogue electrical circuits, such as high-pass filters or resonators. They also used water-based ink made from keratin and graphene to functionalize cellulose, resulting in flexible electrodes with good electrical conductivity even after multiple folding and unfolding cycles. The authors highlighted the circular potential of keratin, referring to it as "wastetronics" – the valorization of waste materials for electronic applications.

The study by Gao et al. (2020) demonstrated the use of hydrolyzed keratin embedded in polyacrylamide to fabricate hydrogel transducers for human–machine interface systems. These hydrogels can detect and measure human movements, breathing patterns, vocal cord vibrations and facial expressions. Incorporating hydrolyzed keratin gave the hydrogel a lower modulus and exceptional extensibility. These hydrogels exhibit high transparency, wide temperature tolerance (-20°C–50°C), stable conductivity, good pressure and strain sensitivity over a wide range and robust adhesion without delamination. These material characteristics offer new opportunities for the design of next-generation skin-like electronic devices, artificial intelligence systems and health monitoring devices (Gao et al., 2020).

Collagen

Collagen is the most abundant protein in nature and is typically found in the extracellular matrix of connective tissues. One of its common derivatives is gelatin, which exhibits robust mechanical properties, including tensile strength and elasticity. Due to their biocompatibility, transparency and insolubility, collagen and gelatin have been found to have applications as substrates in flexible and biodegradable electronics. They have also been used as temperature, deformation and pressure sensors, converting the physical changes into an electrical signal (Edward & Golecki, 2023). In addition, collagen is suitable for creating electronic skins and coating sensors to detect pathogenic microorganisms (Khan & Khan, 2013; Meyer, 2019).

DNA

DNA, or deoxyribonucleic acid, is a biological, self-replicating polymer that carries genetic information in most organisms. DNA has unique properties that make it particularly interesting for producing electronic components. For example, some studies have demonstrated its ability to conduct electricity (Kratochvílová et al., 2008), and its self-assembly sequence can be programmed, making it an intriguing solution for nanoelectronics (Taniguchi & Kawai, 2006).

DNA is extremely interesting as a functional nanoscale material. DNA-based biopolymers have been used to fabricate transistors as an alternative to synthetic methyl methacrylate polymers and to develop high-density memory devices. In particular, DNA-based data storage is increasingly emerging as a promising solution for permanent digital information storage due to its unique biological properties. Condensed DNA has significant potential to become a storage material of the future (Ping et al., 2019). Adleman, a computer science and molecular biology professor at the University of Southern California,[7] is credited with pioneering the conceptual revolution of biological computing. He conducted an experiment that demonstrated the feasibility of computing at the molecular level using DNA. This showed how biological computers could be created using living cells and chemical inputs, including biologically derived molecules such as proteins and DNA, to perform computational tasks involving data storage, retrieval and processing (Adleman, 1994; Parker, 2003). For example, the data density of DNA is 1,000 times higher than that of a solid-state hard disk and 300 times more durable than stable magnetic tape (Rothemund, 2006; Seeman et al., 2017). A new electrogenetic framework for directly storing digital data in living cells used a CRISPR adaptation system, encoding binary data into 3-bit units into CRISPR arrays of bacterial cells via electrical stimulation. This procedure makes significant information storage possible and has a capacity of up to 72 bits, which can be maintained for many generations in open natural environments (Yim et al., 2021).

The potential applications of DNA in electronics cover a wide range of fields. For example, using an editable DNA-traceable barcode label, Liu et al. (2022) have developed a novel method for tracking and verifying food's geographical origin and authenticity. This approach consists of converting traditional digital information into DNA sequences. In order to ensure the stability and detectability of these sequences, the researchers used an encapsulation process that protects the DNA-traceable barcode during storage and allows it to be easily read using DNA sequencing and PCR capillary electrophoresis.

Active Pigments

So far, our research has focused on biobased and biomanufactured materials for various applications as electronic components in potential interactive devices. Biobased pigments, which can be used as electrical conductors and sensors, are another important material class that deserves closer examination. Among the biobased pigments under extensive investigation in this area are melanin, indigo and carotenoids.

Melanins are ubiquitous biopigments responsible for natural skin, hair and eye color. In particular, eumelanin is a low-conductivity type of melanin that produces dark pigmentation in hair, eyes and skin and is known for its limited electrical conductivity. Interestingly, a recent study has shown that when eumelanin is subjected to thermal annealing in a vacuum, its conductivity can be significantly increased to the extent that it could potentially be used as a biocompatible conductive coating agent (Migliaccio et al., 2019). In another study, the potential of this biologically derived material with low human and environmental toxicity was demonstrated using sepia melanin ink films to develop printed electronics (Camus et al., 2022).

Indigo is a natural dye widely recognized as one of the world's best-known and most cost-effective pigments. It is commonly used in paints and cosmetics and as a denim dye. In addition, indigo is an excellent candidate for sustainable electronics based on biodegradable and biocompatible materials due to its highly stable organic semiconductor properties. (Głowacki et al., 2012).

Carotenoids, on the other hand, are organic pigments with semiconducting properties. They are extracted from plants, algae or bacteria. In photosynthetic organisms (and non-photosynthetic organisms), these pigments play a vital role in the light-harvesting photosystems. Due to their molecular structure and strong light-absorbing properties, carotenoids are of great interest for photosensitization and controlled electrical conduction, making them attractive as active materials for organic photovoltaic cells. (Singh & Mukherjee, 2022).

LIVING ORGANISMS

Up to this point in the chapter, the focus has been on biobased materials that can serve as alternatives to commonly used and environmentally harmful materials in IEDs. In the following paragraph, living organisms used in the production of biofabricated materials will be introduced, covering both the use of these materials as inert substrates and technical components and the incorporation of living organisms in interactive applications – or living interactive systems, as illustrated in the next chapter with case studies that have already been developed. Living organisms can offer new opportunities to design interactive devices due to their ability to grow, transform, self-repair, sense and adapt autonomously to their environment. These organisms can be used in their natural state or genetically modified using engineering techniques borrowed from synthetic biology to enhance their inherent properties or to impart programmable functionalities (Stephanopoulos, 2022). Biodesign and biotechnologies are based on integrating living organisms into the manufacturing process to produce specific materials or to harness their ability to sense and interact with their environment, thus incorporating these capabilities into projects. This section explores the capabilities of the most studied organisms in these fields, ranging from constructive capabilities, valuable for the creation of biofabricated materials or architectures, to their sensory and cognitive capabilities, which make living organisms increasingly attractive for future biobased electronics, biocomputing[8] and living interactive systems.

Mycelium

Mycelium, the underground vegetative part of fungi, consisting of an intricate network of white filaments, can bind with other organic elements, resulting in lightweight and compact structures. Once the mycelium has grown, the material is dried to give it stability and durability. Several companies, such as Ecovative[9] in the US and MOGU[10] in the EU, are actively producing mycelium-based products and kits for making mycelium artifacts. They are making significant progress in exploring potential commercial applications.

Mycelium is an excellent and sustainable prototyping material due to its controlled growth and adaptability to shaping and laser engraving. Numerous projects have demonstrated the potential of mycelium for rapid prototyping of low-fidelity electronic and interactive enclosures. Mycelium allows experimentation with intricate three-dimensional shapes by creating molds using digital fabrication systems in which the mycelium can be grown. This opens up a realm of new and unique possibilities for interaction design, resulting in innovative forms of ergonomics and usability. In addition, interactive artifacts can take on a more natural and distinctive character by being integrated into the soft, organic morphologies of the mycelium. Each mycelial artifact has its unique aesthetic qualities, including variations in texture, color and hardness, despite the use of molds. These variations are due to factors such as the mold's shape, the mold's substrate material, and environmental conditions such as humidity, temperature and nutrient levels. This enables the creation of a new generation of unique, irregular interactive artifacts. For example, combining wood, fungal mycelium and cellulose nanofibrils through various processes can result in multiple hybrid panel composites (Sun et al., 2019).

Mycelium can be applied as a substrate for embedding circuits or sensors, as demonstrated by projects such as Myco-Accessories, which emphasizes sustainability in the prototyping process by creating wearables that incorporate biodegradable materials to house electronics. Myco-Accessories exploits the heat resistance, thermal stability and hydrophobic properties of mycelium. This makes it well-suited for wearables. The project promotes a sustainable lifecycle in which the electronics can be reused after wearing the accessory, while the mycelium can be composted (Vasquez & Vega, 2019).

Pure mycelium can also be cultivated in sheets that closely resemble animal leather, particularly in the textile sector. Remarkable work in this field can be seen, for example, in the projects carried out by Mycoworks.[11] In this sheet form, fungal leather has been experimented with as a biodegradable substrate for sustainable electronics, often referred to as "fungal mycelium skin", which serves as a key element of a novel approach known as MycelioTronics (Danninger et al., 2022).

In their living state, fungi possess almost all the sensory mechanisms used by humans and can respond to various chemical and physical stimuli by altering their electrical activity. They can detect and respond to factors such as light, chemicals, gases, gravity, electric fields, changes in substrate pH, mechanical stimulation, toxic metals, CO2 levels and stress hormones (Adamatzky, 2021). These remarkable sensing capabilities open possibilities for applications in future scenarios, such as architecture and wearable technologies, as well as for biological interactive objects.

The cross-disciplinary EU-funded project Fungar[12] is dedicated to the development of a fully integrated structural and computational living substrate using fungal mycelium. The main objective is to cultivate living structures and enhance the functionality of the mycelial network to serve as a computing device. Materials composed of or incorporating mycelium could potentially be transformed into an extensive distributed sensor network. This innovative approach could open up new avenues of research with applications in wearable technology and the Internet of Things (IoT) (Adamatzky, 2021).

Physarum Polycephalum

Slime molds are part of the kingdom Protista. They can be cellular slime molds (formed from multiple cells) or plasmodial slime molds (formed from a large multi-nucleated single cell with thousands of nuclei without a membrane between them). One well-known plasmodial slime mold is *Physarum polycephalum*, a single-celled amoeba-like organism. It is famous for its intelligent behavior. A renowned experiment (Nakagaki et al., 2000) demonstrated its ability to find food by choosing the shortest route in a maze. During the vegetative stage, the *Physarum* transforms into a plasmodium, producing a yellowish vascular network that extends in search of food and makes complex multi-objective foraging decisions (Awad et al., 2023). This organism exhibits distributed sensing, information processing and decision-making capabilities (Adamatzky, 2009). Adamatzky's research positions *Physarum* as a biological model for unconventional computing, suggesting its potential as a programmable biological machine (Adamatzky, 2010). His work explores the space–time dynamics of the plasmodium's response to localized illumination, suggesting its application in illumination-driven computing devices (Adamatzky, 2009).

Two EU-funded projects, PhyChip and PhySense,[13] have investigated *Physarum polycephalum* as a living energy transducer for mechanical, optical and chemical sensors (Adamatzky, 2021). These biosensors hold promise for applications in several fields, including drug discovery, biomedicine, food safety, defense, security and environmental monitoring. Due to its autonomy and resilience in different environments, *Physarum polycephalum* could also find applications in future wearable devices (Adamatzky, 2021).

In addition to its use in sensing and computing, *Physarum polycephalum* has been involved in interactive design projects. The Bio Ornamentum project, conducted in the Hybrid Design Lab, developed a process whereby users interact with the decoration of an archetypal vase inspired by Etruscan archaeology. Users can guide and instruct the organism by managing environmental parameters and the food source to generate specific decorations (Langella & Fiume, 2023).

The unique behavioral model of these organisms can also be used to instruct artificial intelligence (AI). DeepGreen by EcoLogicStudio (Pasquero & Poletto, 2020) exploits the *Physarum polycephalum* model of intelligence to train generative adversarial network (GAN) algorithms (an approach to generative modeling using deep learning methods) to "behave" like a *Physarum polycephalum*, to disfruit its computational abilities and self-organizing behaviors; the trained GAN_Physarum is then suggested as a urban design technique that could couple biological and artificial intelligence in urban design.

Algae

"Algae" is a collective term for various photosynthetic eukaryotic organisms. They vary widely in size, from microscopic species such as chlorella and diatoms to larger species such as giant kelp, which can grow to 65 meters in length. Algae are abundant and fast-growing and can be farmed sustainably without depleting natural resources or causing significant environmental damage. This makes them an attractive alternative to non-renewable resources commonly used in materials and products for interactive design.

From a materials perspective, certain algal-derived polysaccharides, such as alginate and agarose extracted from brown and red algae, have applications in electronics. Alginate has been used to create flexible mechanical and biometric sensors that adhere well to human skin, making them suitable for wearable interactive devices (Zhang et al., 2022a, 2022b). Agarose has been printed with conductive ink to act as an electrode (Verma et al., 2018). Maccagnani et al. (2019) presented a novel method to create flexible and transparent conductive films from brown algae for green electronics. These films, produced by low-power sputtering of ultrathin gold layers onto sodium alginate free-standing films, are thin, easy to handle and exhibit excellent conductivity. This manufacturing process uses less energy, avoids toxic components or chemicals derived from petroleum and allows for environmentally friendly disassembly at the end of its life.

In their living state, algae have unique capabilities that have been explored for various interactive media applications. One such ability is bioluminescence, where certain species of algae can produce light through biochemical reactions. This property has been used to create interactive installations and displays that incorporate living light sources. By manipulating environmental conditions or introducing specific triggers, bioluminescent algae can respond and emit light, creating captivating visual effects (Barati et al., 2021). A team of researchers at the University of California San Diego has developed soft devices embedded with algae that emit bioluminescent light when subjected to mechanical stress. These devices do not require electronics to produce light, making them suitable for building soft robots in low-light environments

(Li et al., 2022). Veron et al. (2013) calculated the amount of light produced by different species of bioluminescent algae; although the light emitted by algae is currently too weak and diffuse for lighting purposes, it is more suitable for communication, signaling, orientation, local guidance and artistic expression.

Dinoflagellates, a type of microscopic algae, are one of the most important groups of marine and freshwater phytoplankton known for their bioluminescence. Pyrofarms, a microalgae production company based in California, is cultivating dinoflagellates, specifically *Pyrocystis fusiformis*, to create a prototype lighting system. This system consists of a transparent tube containing spiral-shaped algae that can produce light for about an hour and a half when equipped with a miniature solar-powered recirculation pump and 12–15 litres of bioluminescent material. The *Pyrocystis* fusiformis inside can be recharged by indirect sunlight (Ofer et al., 2021).

Algae are sensitive to various environmental changes, including light, temperature, mechanical stress, and nutrient levels. This sensitivity can be exploited to develop algae-based biosensors that enable interactive installations to provide real-time feedback and generate dynamic visual or audio experiences (Antonacci & Scognamiglio, 2020; Liao et al., 2020). Certain species of algae can respond to touch or pressure, creating touch-sensitive algae-based interfaces (Maldonado & Latz, 2007). For example, Jason Kazuo Takeuchi (2012) has experimented with touch-sensitive algae-based sheets, demonstrating their potential use in interactive tabletops, consoles, displays and other applications.

Microalgae have the potential to generate bioenergy through photosynthesis, which can be used to power interactive media installations or low-power devices. Algae-powered systems contribute to sustainable and environmentally friendly interactive experiences (Hanafi, 2021).

Bacteria

For centuries, medicine has been actively involved in the fight against bacteria, developing antibiotics and disinfectants. These advances have effectively reduced the risks associated with diseases caused by pathogenic bacteria in humans, animals and plants. In recent decades, however, it has become increasingly clear that beneficial bacteria and their byproducts offer significant potential for improving health and creating new production and processing models more in tune with the delicate balance of ecosystems. In the field of biodesign and biotechnologies, bacteria are being explored for their potential to produce materials ranging from bacterial cellulose to commercial PHA microbial bioplastics. Another intriguing capability of bacteria in materials science is their ability to produce bricks without the need for energy. This process uses bacterial biomineralization, as exemplified by BioMason.[14] This company has already delivered and commercialized Biocement®, a building material made of sand transformed into sandstone by the action of bacteria and nutrients into the sand through microbial activity. The resulting bio-bricks have strength and rigidity comparable to conventional bricks made with cement and hydraulic lime additives (Bernardi et al., 2014).

The ability of bacteria to produce minerals, in particular calcium carbonate crystals, can be controlled by a process known as microbially induced carbonate precipitation (MICP). This technique has been proposed as a solution for the remediation and protection of building materials, focusing on self-healing concrete. Bacterial spores are mixed with nutrients and calcium sources to repair cracks by filling them with bioprecipitates through their metabolic activities (Vermeer et al., 2021). In addition, studies have shown that incorporating genetically modified bacteria into mycelium-based materials can impart impressive smart capabilities, including self-repair, sensing and programmability (McBee et al., 2022).

Bacteria also exhibit remarkable features with potential applications in electronics and interactive devices. A recent study at Sun Yat-Sen University in China has demonstrated the feasibility of programming bacterial populations to self-assemble, resulting in adhesion, self-healing and self-assembly capabilities for use in wearable sensors capable of detecting bioelectric or biomechanical signals. This innovative approach, known as Living Assembled Material by Bacterial Adhesion (LAMBA), involves engineering bacteria with pairs of nanobodies–antigens anchored to their outer membrane. These bacteria are cultured separately and then mixed, facilitating their adhesion to each other and forming functional structures (Chen et al., 2022).

Both native and genetically modified bacteria have applications in the pharmaceutical industry for the production of living therapeutics. This includes the use of viable cells equipped with genetic modules to produce advanced drugs that can prevent or treat diseases in the body in a more biocompatible manner than conventional drugs (Tang et al., 2021). In addition, researchers have developed an ingestible micro-bioelectronic device incorporating encapsulated bacteria and electronic photodetectors to diagnose gastrointestinal diseases (Mimee et al., 2018).

Integrating bacteria into artifacts could also provide visual interaction through color change or luminescence. *Flavobacteria*, commonly found in marine environments, can uniquely form organized colonies that exhibit bright and iridescent colors, making them a potential medium for responsive user interfaces. Groutars et al. presented a tool to capture and characterize these changes, including changes in shape, texture and iridescent color, inspiring human–computer interaction (HCI) designers to explore the temporal qualities of living media for future user interfaces (Groutars et al., 2022).

Like algae, bioluminescent bacteria possess the remarkable ability to generate light, leading to research into the creation of glowing devices that utilize the light communication observed in nature. Researchers are actively investigating the optical properties and biosynthetic pathways of bioluminescent bacterial systems for interactive lighting applications. Some living lighting devices rely on genetically engineered bacteria or hybrid systems in which microorganisms contribute electrons to power lighting elements such as LEDs (Ooms et al., 2022). This emerging field offers a unique opportunity for individuals to cultivate light by caring for bioluminescent organisms as if they were pets, moving away from the conventional act of simply flipping a switch.

SUMMING UP: THE POTENTIAL OF BIODERIVED MATERIALS FOR ELECTRONIC AND INTERACTIVE DEVICES

Research into sustainable and biobased material alternatives is essential to facilitate a transition towards sustainability in interactive and electronic devices. This importance is underlined by the growing number of research centers focusing on this topic, which we will explore in more detail in Chapter 5. Research centers worldwide focus on printed and organic electronics, broadening the scope of the materials in sectors as wide-ranging as the Internet of Things (IoT), packaging, healthcare and construction. This expansion highlights the versatility and potential of these innovative materials to transform multiple industries.

In this chapter, we explored the potential of various living organisms to produce valuable materials that can be used as components in electronic and interactive devices. We also emphasized how these organisms can contribute their sensory and responsive capacities to design disciplines and emerging fields such as bio-human-computer interaction (bio-HCI).

However, two significant gaps were also identified. Firstly, although bio-based materials for electronics show great promise, their practical applications are still limited. Many of the case studies discussed are unique prototypes or are still in the research stage. Secondly, transitioning from experimentation to scalable implementation requires further technological development and interdisciplinary collaboration.

Nevertheless, growing interest in bio-based materials reflects increasing awareness that a more resource-conscious, integrated use of organic matter in engineering and design applications is essential to open new pathways towards a circular, regenerative economy and promote long-term sustainability.

On the other hand, do-it-yourself (DIY) design approaches in electronics have received limited attention. Although the field of human–computer interaction (HCI) has explored DIY techniques for electronics with a growing interest in sustainable materials, the need for specialized equipment and chemicals (which are difficult to obtain and handle outside laboratory facilities) may discourage designers. As a result, experimental cases using a DIY-Materials approach (Rognoli et al., 2015; Rognoli & Ayala-Garcia, 2021) mainly rely on bioplastics (e.g. those derived from starch, algae or animal gut), which may lack inherent electronic functionalities, but can serve as biobased substrates or can be loaded with reactive inks and conductive materials that transform into smart materials.

At the same time, the electronics industry is increasingly responding to the imperative of sustainability with increased research into biobased materials. This includes recognizing the value of a transdisciplinary approach to discovering new and innovative materials applications. Research efforts such as those led by Koelle and colleagues seek to bridge the worlds of electronics, design and DIY approaches. Recognizing the lack of DIY techniques for fabricating functional electronic circuits and sensors, their research in prototyping soft devices with interactive bioplastics

(Koelle et al., 2022) actively demonstrates the potential of organic and easily adaptable materials for electronics, contributing to the field of interactive bioplastics and providing designers with new knowledge and material options for prototyping. At this still experimental research stage, a hybrid professionalism is beginning to emerge, working at the intersection of design and science to develop new sustainable solutions.

The recent rise of hyper-specialized materials such as technopolymers, superconductors and nanomaterials (to name but a few) has distanced designers from their ability to create and shape conventional materials (Langella, 2021). This rupture between design and materials is now leading to a trend where designers are attempting to regain material knowledge through new DIY experiments. European research projects such as MaDe,[15] which sought to bring out the figure of the designer as a designer of materials in the tradition of Italian design, confirm this trend. Designers are increasingly inspired by nature, observing and studying it with the help of scientific tools and discoveries. In the field of material design, there are many reasons why designers are turning to science, especially in the search for sustainable innovations. However, designers don't always act fully aware of scientific processes; therefore, transdisciplinary collaborations are essential for effective design outcomes, and the intersection of design and science needs to be based on shared goals and language (Langella, 2021). Alongside a growing number of designers who approach biodesign by equipping domestic spaces and design studios to work with living organisms, an even more niche figure emerges in this panorama: the designer working in a scientific laboratory. Domestic and design laboratories are non-specialized spaces that do not have access to the same scientific laboratory tools that guarantee high-tech processes, precise measurements and a high level of cleanliness to limit contamination; this can limit the type of research that can take place and the type of organisms that can be addressed (Crawford, 2023). With the definition of "designer in lab" (Langella, 2021), we refer to a new type of designer who personally experiences a direct relationship with the project of matter and who carries out their activities in scientific laboratories and no longer only in the design studio. Such designers must embrace the scientific method, the rigor of experimentation, protocols and rational criteria, hybridizing their discipline to the extent that they move to the scientific laboratory as a working environment. In some universities and research centers, this transdisciplinary approach is also supported by facilitating interaction between design and other life science disciplines (some examples are MIT in Boston, the Hub for Biotechnology in the Built Environment in Newcastle, the Bio-Integrated Design Lab in London, and the Biodesign Lab of TU Delft); however, it can still be a challenge for designers to enter scientific laboratories. However, such transdisciplinary collaborations can be beneficial for both design and science. Design can help translate the hyper-technicalities of science for the general public and facilitate the vision of new applications of materials or even contribute to the act of discovery; on the other hand, design is empowered by scientific knowledge and its tools (Pollini, 2023). Even if the examples of designers working in laboratories can still be considered isolated experiences, just as biodesign is still an emerging discipline, this path will likely continue and evolve as the biotechnology sector expands in parallel. Even the field of interaction design, as we will see in the following chapters, is positively influenced by this trend, especially in the case of interactions with living organisms, where a scientific approach and working environment are often the only possible circumstances for such interactive and material experiments.

NOTES

1 The CHEMARTS Cookbook is an open-access publication by Aalto University that compiles recipes and experiments developed by students and researchers to explore biomaterials from natural polymers such as cellulose. It serves as an educational tool to promote

hands-on learning and innovation in sustainable materials. Retrieved in March 2025 from: https://chemarts.aalto.fi/index.php/the_chemarts_cookbook/.

2 Professor Orlando Rojas Research Group, Biobased Colloids and Materials; retrieved Marche 2025 from: https://rojas.chbe.ubc.ca/ The Bioproduct Institute at University of British Columbia; Retrieved March, 2025, from https://bpi.ubc.ca/

3 he Self-Assembly Lab at MIT is a cross-disciplinary research group exploring programmable materials and adaptive systems through the convergence of design, architecture, science, and engineering. It is known for pioneering work on materials that can change shape, assemble, or respond to environmental stimuli. Retrieved in March 2025 from: https://selfassemblylab.mit.edu/.

4 The InnoCell project, developed by the interdisciplinary Design Friction Lab at the Free University of Bozen-Bolzano, explores the potential of bacterial cellulose in both edible and non-edible design applications, combining speculative design with scientific research. The project is based on an expired US patent describing a rotating disc bioreactor system for microbial cellulose production, originally developed by Bungay III and Serafica. See: Bungay III, H.R., & Serafica, G.C. (1999). Production of microbial cellulose using a rotating disc film bioreactor. US Patent No. 5,955,326. Retrieved March, 2025, from https://designfrictionlab.com/project/innocell-mod/

5 DeForma is a research project coordinated by Politecnico di Milano, focused on exploring the potential of bacterial cellulose in material design. The project investigates biotinkering practices, hybridization with other materials, and applications in wearable technology and healthcare. Retrieved on March 2025 from: https://www.deforma.polimi.it/

6 This definition describes two types of bacteria distinguished by Gram stain, a staining method used to classify bacterial species into two large groups that differ in chemical and physical properties: gram-positive bacteria retain purple crystals and stain purple thanks to their thick cell walls, while gram-negative bacteria lose them red coloration due to thinner cell wall.

7 Leonard Adleman is a professor of computer science and molecular biology at the University of Southern California (USC) Viterbi School of Engineering. He is widely known for his pioneering work in DNA computing. Retrieved March, 2025, from https://viterbi.usc.edu/directory/faculty/Adleman/Leonard

8 For the purposes of this book, Biocomputing is defined as Biological Computing, namely the use of biologically derived components and materials (e.g., DNA) to perform computational functions. See also Merriam-Webster definition of Biocomputer; Retrieved March, 2025, from https://www.merriam-webster.com/dictionary/biocomputer

9 Ecovative is a U.S.-based company specializing in the development and commercialization of mycelium-based technologies for packaging, textiles, food, and construction materials. Retrieved March, 2025, from https://www.ecovative.com/

10 Mogu is an Italian company focused on sustainable mycelium-based materials for interior design and architecture, offering acoustic panels, flooring solutions, and other biofabricated products. Retrieved March, 2025, from https://mogu.bio/
11 MycoWorks is a U.S.-based biotechnology company that engineers mycelium to create Reishi™, a fine mycelium material developed as a sustainable and high-performance alternative to leather. Retrieved March, 2025, from https://www.mycoworks.com/
12 FUNGAR (Fungal Architectures) is an EU-funded research project exploring the use of fungal mycelium as a living material for computational and architectural applications, including biocomputing and smart sensing systems. Retrieved March, 2025, from https://www.fungar.eu/
13 SUCCESS (SUrface wettability ConCEptS for advanced interfaceS) is a research project funded by the European Union focusing on advanced surface technologies inspired by biological systems. It aims to develop innovative materials and surfaces with specific wettability characteristics for applications in health, energy, and environment.
14 BioMason is a biotechnology company that develops cement-free construction materials through microbial biocementation processes. By using naturally occurring bacteria to grow strong, durable materials, BioMason aims to significantly reduce carbon emissions in the construction industry. Retrieved March, 2025, from https://biomason.com/
15 MaDe (Material Designers, 2020) is a project funded by the Creative Europe programme and led by Elisava, Politecnico di Milano and Matt-er London. The project promotes material design as a driver for circular innovation, connecting designers, researchers and industry. Through exhibitions, workshops and awards, MaDe supports emerging talent exploring sustainable material futures. MaDe – Material Designers. (2020). Material Designers: A Creative Europe project fostering material-led innovation for a circular economy. Elisava Barcelona School of Design and Engineering, Politecnico di Milano & Ma-tt-er London. Co-funded by Creative Europe. Retrieved from https://research.elisava.net/assets/documents/material-designers.pdf

REFERENCES

Abascal, N. C., & Regan, L. (2018). The past, present and future of protein-based materials. *Open Biology*, 8(10), 180113. https://doi.org/10.1098/rsob.180113

Adamatzky, A. (2009). Steering plasmodium with light: Dynamical programming of Physarum machine. *New Mathematics and Natural Computation Journal*. Advance online publication. Retrieved March, 2025, from http://arxiv.org/abs/0908.0850

Adamatzky, A. (2010). *Physarum machines: Computers from slime mould*. World Scientific. https://doi.org/10.1142/7968

Adamatzky, A., Nikolaidou, A., Gandia, A., & Chiolerio, A. (2021). Living wearables from slime mould and fungi. *Revue LINKs series Special Issue, 1*, 93–98.

Adleman, L. M. (1994). Molecular computation of solutions to combinatorial problems. *Science (New York, N.Y.), 266*(5187), 1021–1024. https://doi.org/10.1126/science.7973651

Altaf, A., Usmani, Z., Dar, A. H., & Dash, K. K. (2022). A comprehensive review of polysaccharide-based bionanocomposites for food packaging applications. *Discover Food, 2*(1), 10. https://doi.org/10.1007/s44187-022-00011-x

Antonacci, A., & Scognamiglio, V. (2020). Biotechnological advances in the design of algae-based biosensors. *Trends in Biotechnology, 38*(3), 334–347. https://doi.org/10.1016/j.tibtech.2019.10.005

Awad, A., Pang, W., Lusseau, D., & Coghill, G. M. (2023). A survey on physarum polycephalum intelligent foraging behaviour and bio-inspired applications. *Artificial Intelligence Review, 56*(1), 1–26. https://doi.org/10.1007/s10462-021-10112-1

Azeredo, H. M. C., Barud, H., Farinas, C. S., Vasconcellos, V. M., & Claro, A. M. (2019). Bacterial cellulose as a raw material for food and food packaging applications. *Frontiers in Sustainable Food Systems, 3*, Article 7. https://doi.org/10.3389/fsufs.2019.00007

Barati, B., Karana, E., Pont, S., & Van Dortmont, T. (2021, June). Living light interfaces–an exploration of bioluminescence aesthetics. In *Designing interactive systems conference 2021* (pp. 1215–1229). https://doi.org/10.1145/3461778.3462038

Bell, F., Chow, D., Lazaro, E., Devendorf, L., & Alistar, M. (2023). Designing interactions with Kombucha Scoby. In *TEI '23: Seventeenth international conference on tangible, embedded, and embodied interaction*, 5. https://doi.org/10.1145/3569009.3571841

Bell, F., Coffie, J., & Alistar, M. (2024). Bio-digital calendar: Attuning to non-human temporalities for multispecies understanding. In *Proceedings of the eighteenth international conference on Tangible, Embedded, and Embodied Interaction (TEI' 24)* (Article 38, pp. 1–15). Association for Computing Machinery. https://doi.org/10.1145/3623509.3633386

Bernardi, D., DeJong, J. T., Montoya, B. M., & Martinez, B. C. (2014). Bio-bricks: Biologically cemented sandstone bricks. *Construction and Building Materials, 55*, 462–469. https://doi.org/10.1016/j.conbuildmat.2014.01.019

Blanco, A., Monte, M. C., Campano, C., Balea, A., Merayo, N., & Negro, C. (2018). Nanocellulose for industrial use: Cellulose Nanofibers (CNF), Cellulose Nanocrystals (CNC), and Bacterial Cellulose (BC). In *Handbook of nanomaterials for industrial applications* (pp. 74–126). Elsevier. https://doi.org/10.1016/B978-0-12-813351-4.00005-5

Bungay, H. R., III, et al. (1999). Production of microbial cellulose using a rotating disk film bioreactor (U.S. Patent No. 5,955,326). U.S. Patent and Trademark Office Routledge. https://patents.google.com/patent/US5955326A

Camere, S., & Karana, E. (2017). Growing materials for product design. *International Journal of Design*, *11*(3), 1–14.

Cataldi, P., Condurache, O., Spirito, D., Krahne, R., Bayer, I. S., Athanassiou, A., & Perotto, G. (2019). Keratin-graphene nanocomposite: Transformation of waste wool in electronic devices. *ACS Sustainable Chemistry & Engineering*, *7*(14), 12544–12551. https://doi.org/10.1021/acssuschemeng.9b02415

Camus, A., Reali, M., Rozel, M., Zhuldybina, M., Soavi, F., & Santato, C. (2022). High conductivity Sepia melanin ink films for environmentally benign printed electronics. *Proceedings of the National Academy of Sciences*, *119*(32), e2200058119. https://doi.org/10.1073/pnas.2200058119

Chen, B., Kang, W., Sun, J., Zhu, R., Yu, Y., Xia, A., & Dai, Z. (2022). Programmable living assembly of materials by bacterial adhesion. *Nature Chemical Biology*, *18*(3), 289–294.

Clèries, L., Rognoli, V., Solanki, S., & Llorach, P. (Eds.). (2020). Material designers: *A Creative Europe project fostering material-led innovation for a circular economy*. Elisava Barcelona School of Design and Engineering. Retrieved from https://research.elisava.net/assets/documents/material-designers.pdf

Cohen, N., Sicher, E., Ayala-Garcia, C., Sanchez-Fayos, I. M., Conterno, L., & Ugur Yavuz, S. (2023). Innocell bioreactor: An open-source development to produce biomaterials for food and packaging based on fermentation processes. *Fermentation*, *9*(10), 915.

Cohen, N., Sicher, E., & Uğur Yavuz, S. (2019, November). From agricultural waste to microbial growth and (G) local resilience. In *UNIDCOM/IADE international conference senses & sensibility* (pp. 81–92). Springer International Publishing.

Crawford, A. (2023). *Designer's guide to lab practice*. Routledge. https://doi.org/10.4324/9781003363774

Danninger, D., Pruckner, R., Holzinger, L., Koeppe, R., & Kaltenbrunner, M. (2022). MycelioTronics: Fungal mycelium skin for sustainable electronics. *Science Advances*, *8*(45), eadd7118. https://doi.org/10.1126/sciadv.add7118

Eder, M., Amini, S., & Fratzl, P. (2018). Biological composites—complex structures for functional diversity. *Science*, *362*(6414), 543–547. https://doi.org/10.1126/science.aat8297

Edward, S., & Golecki, H. M. (2023). Gelatin soft actuators: Benefits and opportunities. *Actuators*, *12*(2), Article 2. https://doi.org/10.3390/act12020063

Erb, R. M., Sander, J. S., Grisch, R., & Studart, A. R. (2013). Self-shaping composites with programmable bioinspired microstructures. *Nature Communications*, *4*(1), Article 1. https://doi.org/10.1038/ncomms2666

Ferrarese, F. M., Pollini, B., Luzio, A., & Caironi, M. (2023, under evaluation). Bacterial cellulose from Kombucha's SCOBY as multipurpose material for fully edible electronics. *European-MRS 2023*, Strasbourg.

Fontana, S. S. (2023). *Discovering KBC: Manufacturing techniques with kombucha bacterial cellulose* [Master thesis, Politecnico di Milano]. https://www.politesi.polimi.it/handle/10589/201402

Fratzl, P. (2021). *Lectures on the interdependence of structure and function in biological materials*. Matters of Activity. Image Space Material. Retrieved March 2025, from https://www.matters-of-activity.de

Gallegos, A. M. A., Carrera, S. H., Parra, R., Keshavarz, T., & Iqbal, H. M. (2016). Bacterial cellulose: A sustainable source to develop value-added products–A review. *BioResources*, *11*(2), 5641–5655. http://ncsu.edu/bioresources

Gao, Y., Gu, S., Jia, F., Wang, Q., & Gao, G. (2020). "All-in-one" hhydrolysedkeratin protein-modified polyacrylamide composite hydrogel transducer. *Chemical Engineering Journal*, *398*, 125555.

Ginsberg, A. D., & Chieza, N. (2018). Editorial: Other biological futures. *Journal of Design and Science*. (4). https://doi.org/10.21428/566868b5

Głowacki, E., Voss, G., Leonat, L., Irimia-Vladu, M., Bauer, S., & Sariciftci, N. (2012). Indigo and Tyrian purple – From ancient natural dyes to modern organic semiconductors. Israel Journal of Chemistry, 52(6), 540–551. https://doi.org/10.1002/chin.201244249

Groutars, E. G., Risseeuw, C. C., Ingham, C., Hamidjaja, R., Elkhuizen, W. S., Pont, S. C., & Karana, E. (2022, April). Flavorium: An exploration of Flavobacteria's living aesthetics for living color interfaces. In *Proceedings of the 2022 CHI conference on human factors in computing systems* (pp. 1–19).

Groll, J., Boland, T., Blunk, T., Burdick, J. A., Cho, D.-W., Dalton, P. D., Derby, B., Forgacs, G., Li, Q., Mironov, V. A., Moroni, L., Nakamura, M., Shu, W., Takeuchi, S., Vozzi, G., Woodfield, T. B. F., Xu, T., Yoo, J. J., & Malda, J. (2016). Biofabrication: Reappraising the definition of an evolving field. *Biofabrication*, *8*(1), 013001. https://doi.org/10.1088/1758-5090/8/1/013001

Guidetti, G., d'Amone, L., Kim, T., Matzeu, G., Mogas-Soldevila, L., Napier, B., Ostrovsky-Snider, N., Roshko, J., Ruggeri, E., & Omenetto, F. G. (2022). Silk materials at the convergence of science, sustainability, healthcare, and technology. *Applied Physics Reviews*, 9(1), 011302. https://doi.org/10.1063/5.0060344

Hanafi, W. H. H. (2021). Bio-algae: A study of an interactive facade for commercial buildings in populated cities. *Journal of Engineering and*

Applied Science, *68*(1), 1–16. https://doi.org/10.1186/s44147-021-00037-5
Harrington, M. J., & Fratzl, P. (2021). Natural load-bearing protein materials. *Progress in Materials Science*, *120*, 100767. https://doi.org/10.1016/j.pmatsci.2020.100767
Hosseini, H., Mousavi, S. M., Wurm, F. R., & Goodarzi, V. (2021). Display of hidden properties of flexible aerogel based on bacterial cellulose/polyaniline nanocomposites with help of multiscale modeling. *European Polymer Journal*, *146*, 110251.
Indrarti, L. (2017, March). Incorporation of citrus essential oils into bacterial cellulose-based edible films and assessment of their physical properties. In *IOP conference series: Earth and environmental science* (Vol. 60, No. 1, p. 012018). IOP Publishing. https://doi.org/10.1088/1755-1315/60/1/012018
Interdependence of Structure and Function in Biological Materials—Matters of Activity. (n.d.). Retrieved March, 2025, from https://www.matters-of-activity.de/en/activities/4383/interdependence-of-structure-and-function-in-biological-materials
Jang, W. D., Hwang, J. H., Kim, H. U., Ryu, J. Y., & Lee, S. Y. (2017). Bacterial cellulose as an example product for sustainable production and consumption. *Microbial Biotechnology*, *10*(5), 1181–1185. https://doi.org/10.1111/1751-7915.12744
Junka, A., Żywicka, A., Chodaczek, G., Dziadas, M., Czajkowska, J., Duda-Madej, A., Bartoszewicz, M., Mikołajewicz, K., Krasowski, G., Szymczyk, P., & Fijałkowski, K. (2019). Potential of biocellulose carrier impregnated with essential oils to fight against biofilms formed on hydroxyapatite. *Scientific Reports*, *9*(1), Article 1. https://doi.org/10.1038/s41598-018-37628-x
Kang, S. K., Yin, L., & Bettinger, C. (2020). The emergence of transient electronic devices. *MRS Bulletin*, *45*(2), 87–95. https://doi.org/10.1557/mrs.2020.19
Karana, E. (2020). *Still alive: Livingness as a material quality in design*. Avans University of Applied Sciences.
Karana, E., Barati, B., & Giaccardi, E. (2020). Living artifacts: Conceptualizing livingness as a material quality in everyday artifacts. *International Journal of Design*, *14*, 37–53.
Khan, R., & Khan, M. H. (2013). Use of collagen as a biomaterial: An update. *Journal of Indian Society of Periodontology*, *17*(4), 539. https://doi.org/10.4103/0972-124X.118333
Kim, H. J., Yim, E. C., Kim, J. H., Kim, S. J., Park, J. Y., & Oh, I. K. (2017). Bacterial nano-cellulose triboelectric nanogenerator. *Nano Energy*, *33*, 130–137.
Koelle, M., Nicolae, M., Nittala, A. S., Teyssier, M., & Steimle, J. (2022). Prototyping soft devices with interactive bioplastics. In Proceedings

of the 35th annual ACM symposium on User Interface Software and Technology (UIST '22) (pp. 1–16). ACM.

Kratochvílová, I., Král, K., Bunček, M., Víšková, A., Nešpůrek, S., Kochalska, A., Todorciuc, T., Weiter, M., & Schneider, B. (2008). Conductivity of natural and modified DNA measured by scanning tunneling microscopy. The effect of sequence, charge and stacking. *Biophysical Chemistry*, *138*(1), 3–10. https://doi.org/10.1016/j.bpc.2008.08.005

Langella, C. (2021). Design and science: A pathway for material design. In O. Pedgley, V. Rognoli, & E. Karana (Eds.), *Materials experience 2: Expanding territories of materials and design* (pp. 259–277). Butterworth-Heinemann: Elsevier.

Langella, C., & Fiume, M. (2023). Biodiseño: El proyecto que educa a la naturaleza. *Cuadernos del Centro de Estudios de Diseño y Comunicación*, *203*, 17–28.

Lee, S. (2005). *Fashioning the future: Tomorrow's wardrobe*. Thames and Hudson.

Lee, S., Congdon, A., Parker, G., & Borst, C. (2020). *Understanding 'bio' material innovation: A primer for the fashion industry*. ©Biofabricate and Fashion for Good 2021. https://it1lib.org/book/11722399/3da5ef

Li, C., He, Q., Wang, Y., Wang, Z., Wang, Z., Annapooranan, R., ... Cai, S. (2022). Highly robust and soft biohybrid mechanoluminescence for optical signaling and illumination. *Nature Communications*, 13(1), Article 3914. https://doi.org/10.1038/s41467-022-31705-6

Li, T., Chen, C., Brozena, A. H., Zhu, J. Y., Xu, L., Driemeier, C., Dai, J., Rojas, O. J., Isogai, A., Wågberg, L., & Hu, L. (2021). Developing fibrillated cellulose as a sustainable technological material. *Nature*, *590*(7844), Article 7844. https://doi.org/10.1038/s41586-020-03167-7

Lin, Q., Hao, S., Hu, W., Wang, M., Zang, Z., Zhu, L., & Tang, X. (2019). Human hair keratin for physically transient resistive switching memory devices. *Journal of Materials Chemistry C*, *7*(11), 3315–3321. https://doi.org/10.1039/c8tc05334k

Lin, Y.-H., Kang, P.-L., Xin, W., Yen, C.-S., Hwang, L.-C., Chen, C.-J., Liu, J.-T., & Chang, S. J. (2018). Preparation and evaluation of chitosan biocompatible electronic skin. *Computers in Industry*, *100*, 1–6. https://doi.org/10.1016/j.compind.2018.03.040

Liu, K., Xing, R., Sun, R., Ge, Y., & Chen, Y. (2022). An accurate and rapid way for identifying food geographical origin and authenticity: editable DNA-traceable barcode. *Foods*, *12*(1), 17. https://doi.org/10.3390/foods12010017

Lu, Q., Wang, X., Hu, X., Cebe, P., Omenetto, F., & Kaplan, D. L. (2010). Stabilisationand release of enzymes from silk films. *Macromolecular Bioscience*, *10*(4), 359–368. https://doi.org/10.1002/mabi.200900388

Maccagnani, P., Bertoldo, M., Dinelli, F., Murgia, M., Summonte, C., Ortolani, L., & Capelli, R. (2019). Flexible conductors from brown algae for green electronics. *Advanced Sustainable Systems*, *3*(6), 1900001. https://doi.org/10.1002/adsu.201900001
Maldonado, E. M., & Latz, M. I. (2007). Shear-stress dependence of dinoflagellate bioluminescence. *The Biological Bulletin*, *212*(3), 242–249. https://doi.org/10.2307/25066606
Malik, H., Niazi, M. B. K., Miran, W., Tawfeek, A. M., Jahan, Z., Kamel, E. M., Ahmed, N., & Saeed Akhtar, M. (2023). Wood as a green and sustainable alternative for environmentally friendly & flexible electronic devices. *Chemosphere*, *336*, 139213. https://doi.org/10.1016/j.chemosphere.2023.139213
Marelli, B., Brenckle, M. A., Kaplan, D. L., & Omenetto, F. G. (2016). Silk fibroin as edible coating for perishable food preservation. *Scientific Reports*, *6*(1), Article 1. https://doi.org/10.1038/srep25263
Matzeu, G., Mogas-Soldevila, L., Li, W., Naidu, A., Turner, T. H., Gu, R., Blumeris, P. R., Song, P., Pascal, D. G., Guidetti, G., Li, M., & Omenetto, F. G. (2020). Large-scale patterning of reactive surfaces for wearable and environmentally deployable sensors. *Advanced Materials*, *32*(28), 2001258. https://doi.org/10.1002/adma.202001258
McBee, R. M., Lucht, M., Mukhitov, N., Richardson, M., Srinivasan, T., Meng, D., & Wang, H. H. (2022). Engineering living and regenerative fungal-bacterial biocomposite structures. *Nature Materials*, *21*(4), 471–478.
Merritt, T., Hamidi, F., Alistar, M., & DeMenezes, M. (2020). Living media interfaces: A multi-perspective analysis of biological materials for interaction. *Digital Creativity*, *31*(1), 1–21. https://doi.org/10.1080/14626268.2019.1707231
Meyer, M. (2019). Processing of collagen-based biomaterials and the resulting materials properties. *Biomedical Engineering Online*, *18*(1), 1–74. https://doi.org/10.1186/s12938-019-0647-0
Migliaccio, L., Manini, P., Altamura, D., Giannini, C., Tassini, P., Maglione, M. G., Minarini, C., & Pezzella, A. (2019). Evidence of unprecedented high electronic conductivity in mammalian pigment-based eumelanin thin films after thermal annealing in vacuum. *Frontiers in Chemistry*, *7*, 162. https://www.frontiersin.org/articles/10.3389/fchem.2019.00162
Mimee, M., Nadeau, P., Hayward, A., Carim, S., Flanagan, S., Jerger, L., Collins, J., McDonnell, S., Swartwout, R., Citorik, R. J., Bulović, V., Langer, R., Traverso, G., & Lu, T. K. (2018). An ingestible bacterial-electronic system to monitor gastrointestinal health in situ. *Science*, *360*(6391), 915–918. https://doi.org/10.1126/science.aas9315
Mironov, V., Trusk, T., Kasyanov, V., Little, S., Swaja, R., & Markwald, R. (2009). Biofabrication: A 21st-century manufacturing paradigm. *Biofabrication*, *1*, 022001. https://doi.org/10.1088/1758-5082/1/2/022001

Myers, W. (2012). *Bio design: Nature, science, creativity*. MoMA and Thames & Hudson.

Mo, C., Xiang, L., & Chen, Y. (2021). Advances in injectable and self-healing polysaccharide hydrogel based on the Schiff base reaction. *Macromolecular Rapid Communications*, *42*(10), 2100025. https://doi.org/10.1002/marc.202100025

Nakagaki, T., Yamada, H., & Tóth, Á. (2000). Maze-solving by an amoeboid organism. *Nature*, *407*(6803), 470. https://doi.org/10.1038/35035159

Naomi, R., Bt Hj Idrus, R., & Fauzi, M. B. (2020). Plant- vs. bacterial-derived cellulose for wound healing: A review. *International Journal of Environmental Research and Public Health*, *17*(18), 6803. https://doi.org/10.3390/ijerph17186803

Ofer, N., Bell, F., & Alistar, M. (2021). Designing direct interactions with bioluminescent algae. In *Designing interactive systems conference 2021* (pp. 1230–1241). https://doi.org/10.1145/3461778.3462090

Omenetto, F. G. (2017). Silk materials at the intersection of technology and biology. In G. G. Malliaras, T. M. Squires, & S. R. Nagel (Eds.), *Active matter* (pp. 47–50). MIT Press.

Omenetto, F. G., & Kaplan, D. L. (2008). A new route for silk. *Nature Photonics*, *2*(11), Article 11. https://doi.org/10.1038/nphoton.2008.207

Onggo, D., Mulyani, I., Valverde-Muñoz, F. J., Real, J. A., & Molnar, G. (2017). Bistable thermo-chromic and magnetic spin crossover microcrystals embedded in nata de coco bacterial cellulose biofilm. *Cellulose*, *24*(5), 2205–2213.

Ooms, D., Barati, B., Bruns, M., & Van Dongen, T. (2022, October). From concern to care: A transformative reflection on designing-with the living. In Proceedings of the Nordic Human–Computer Interaction Conference (NordiCHI '22) (pp. 1–15). Association for Computing Machinery. https://doi.org/10.1145/3546155.3546666

Papadopoulou, A., Laucks, J., & Tibbits, S. (2017). General principles for programming material. In G. G. Malliaras, T. M. Squires, & S. R. Nagel (Eds.), *Active matter*. MIT Press.

Papile, F., Bolzan, P., Parisi, S., & Pollini, B. (2023). Perceiving grown bacterial cellulose. In E. Duarte & A. Di Roma (Eds.), *Developments in design research and practice II* (pp. 333–346). Springer Nature Switzerland. https://doi.org/10.1007/978-3-031-32280-8_24

Pasquero, C., & Poletto, M. (2020). Deep green. In *ACADIA 2020: Distributed proximities / Volume I: Technical papers [Proceedings of the 40th Annual Conference of the Association of Computer Aided Design in Architecture (ACADIA)*.

Pataranutaporn, P., Ingalls, T., & Finn, E. (2018). Biological HCI: Towards integrative interfaces between people, computer, and biological materials. https://doi.org/10.1145/3170427.3188662

Parker, J. (2003). Computing with DNA. *EMBO Reports*, *4*(1), 7–10. https://doi.org/10.1038/sj.embor.embor719

Parkes, A., & Dickie, C. (2013). A biological imperative for interaction design. In CHI '13 extended abstracts on human factors in computing systems (pp. 2209–2218). Association for Computing Machinery. https://doi.org/10.1145/2468356.2468742

Persin, Z., Stana-Kleinschek, K., Foster, T. J., van Dam, J. E. G., Boeriu, C. G., & Navard, P. (2011). Challenges and opportunities in polysaccharides research and technology: The EPNOE views for the next decade in the areas of materials, food and health care. *Carbohydrate Polymers*, *84*(1), 22–32. https://doi.org/10.1016/j.carbpol.2010.11.044

Ping, Z., Ma, D., Huang, X., Chen, S., Liu, L., Guo, F., Zhu, S., & Shen, Y. (2019). Carbon-based archiving: Current progress and future prospects of DNA-based data storage. *GigaScience*, *8*(5), giz075. https://doi.org/10.1093/gigascience/giz075

Pradhan, S., Brooks, A. K., & Yadavalli, V. K. (2020). Nature-derived materials for the fabrication of functional biodevices. *Materials Today Bio*, *7*, 100065. https://doi.org/10.1016/j.mtbio.2020.100065

Poddar, M. K., & Dikshit, P. K. (2021). Recent developments in bacterial cellulose production and synthesis of cellulose-based conductive polymer nanocomposites. *Nano Select*, *2*(9), 1605–1628.

Pollini, B. (2021). Project 13—Healing materialities from a biodesign perspective. In O. Pedgley, V. Rognoli, & E. Karana (Eds.), *Materials experience 2* (pp. 167–170). Butterworth-Heinemann. https://doi.org/10.1016/B978-0-12-819244-3.00008-9

Pollini, B. (2023). *Healing materialities from a biodesign perspective. Framing biodesigned materials and artifacts for the sustainability transition* [Doctoral thesis, Politecnico di Milano].

Poppinga, S., Correa Zuluaga, D., Bruchmann, B., Menges, A., & Speck, T. (2020). Plant movements as concept generators for the development of biomimetic compliant mechanisms. *Integrative and Comparative Biology*, *60*(4), 886–895. https://doi.org/10.1093/icb/icaa028

Prilepskii, A., Nikolaev, V., & Klaving, A. (2023). Conductive bacterial cellulose: From drug delivery to flexible electronics. *Carbohydrate Polymers*, *313*, 120850. https://doi.org/10.1016/j.carbpol.2023.120850

Rasmussen, S., Chen, L., Deamer, D., Krakauer, D. C., Packard, N. H., Stadler, P. F., & Bedau, M. A. (2004). Transitions from nonliving to living matter. *Science*, *303*(5660), 963–965. https://doi.org/10.1126/science.1093669

Reyssat, E., & Mahadevan, L. (2009). Hygromorphs: From pine cones to biomimetic bilayers. *Journal of The Royal Society Interface*, *6*(39), 951–957. https://doi.org/10.1098/rsif.2009.0184

Revin, V. V., Liyas'kina, E. V., Sapunova, N. B., Shutova, V. V., Vlasov, S. V., & Revin, V. D. (2020). Isolation and ccharacterisationof the strains producing bacterial cellulose. *Microbiology*, *89*, 86–95. https://doi.org/10.1134/S0026261720010130

Rognoli, V., & Ayala-Garcia, C. (2021). Defining the DIY-materials approach. In O. Pedgley, V. Rognoli, E. Karana (Eds.), *Materials experience 2: Expanding territories of materials and design* (pp. 227–258). Butterworth-Heinemann: Elsevier. https://doi.org/10.1016/B978-0-12-819244-3.00010-7

Rognoli, V., Bianchini, M., Maffei, S., & Karana, E. (2015). DIY materials. *Materials and Design*, *86*, 692–702. https://doi.org/10.1016/j.matdes.2015.07.020

Rothemund, P. W. (2006). Folding DNA to create nanoscale shapes and patterns. *Nature*, *440*(7082), 297–302. https://doi.org/10.1038/nature04586

Seeman, N. C., & Sleiman, H. F. (2017). DNA nanotechnology. *Nature Reviews Materials*, *3*(1), 1–23. https://doi.org/10.1038/natrevmats.2017.68

Sicher, E. (2017). *From peel to peel: An experimental design approach for packaging* [BA thesis, Bozen-Bolzano: UNIBZ]. Retrieved March, 2025, from https://unibz.primo.exlibrisgroup.com/permalink/39UBZ_INST/hl3dt9/alma99221147287401241

Singh, A., & Mukherjee, T. (2022). Application of carotenoids in sustainable energy and green electronics. *Materials Advances*, *3*(3), 1341–1358. https://doi.org/10.1039/D1MA01070K

Singhania, R. R., Patel, A. K., Tsai, M.-L., Chen, C.-W., & Di Dong, C. (2021). Genetic modification for enhancing bacterial cellulose production and its applications. *Bioengineered*, *12*(1), 6793–6807. https://doi.org/10.1080/21655979.2021.1968989

Souza, P. R., de Oliveira, A. C., Vilsinski, B. H., Kipper, M. J., & Martins, A. F. (2021). Polysaccharide-based materials created by physical processes: From preparation to biomedical applications. *Pharmaceutics*, *13*(5), 621. https://doi.org/10.3390/pharmaceutics13050621

Stephanopoulos, O. C. (2022). Materials come alive. *Nature Materials*, *21*, 379. https://doi.org/10.1038/s41563-022-01234-0

Sun, W., Tajvidi, M., Hunt, C. G., McIntyre, G., & Gardner, D. J. (2019). Fully bio-based hybrid composites made of wood, fungal mycelium, and cellulose nanofibrils. *Scientific Reports*, *9*(1), 1–12.

Sun, H.-S., Chiu, Y.-C., & Chen, W.-C. (2017). Renewable polymeric materials for electronic applications. *Polymer Journal*, *49*(1), Article 1. https://doi.org/10.1038/pj.2016.95

Takeuchi, J. (2012). Nocturnal light: Exploring the perceptual experience of bioluminescence. http://hdl.handle.net/10125/45687

Taniguchi, M., & Kawai, T. (2006). DNA electronics. *Physica E: Low-Dimensional Systems and Nanostructures*, *33*(1), 1–12. https://doi.org/10.1016/j.physe.2006.01.005

Tang, T. C., An, B., Huang, Y., Vasikaran, S., Wang, Y., Jiang, X., & Zhong, C. (2021). Materials design by synthetic biology. *Nature Reviews Materials*, *6*(4), 332–350.

Tao, H., Hwang, S.-W., Marelli, B., An, B., Moreau, J. E., Yang, M., Brenckle, M. A., Kim, S., Kaplan, D. L., Rogers, J. A., & Omenetto, F. G. (2014). Silk-based resorbable electronic devices for remotely controlled therapy and in vivo infection abatement. *Proceedings of the National Academy of Sciences of the United States of America*, *111*(49), 17385–17389. https://doi.org/10.1073/pnas.1407743111

Udayan, A., Arumugam, M., & Pandey, A. (2017). Nutraceuticals from algae and cyanobacteria. In R. P. Rastogi, D. Madamwar, & A. Pandey (Eds.), *Algal green chemistry* (pp. 65–89). Elsevier. https://doi.org/10.1016/B978-0-444-63784-0.00004-7

Verma, S., Tripathi, J., Katiyar, M., & Verma, V. (2018). Biodegradable photolithography compatible substrate for transparent transient electronics and flexible energy storage devices. *Applied Materials Today*, *13*, 83–90. https://doi.org/10.1016/j.apmt.2018.08.010

Vasquez, E. S. L., & Vega, K. (2019). Myco-accessories: Sustainable wearables with biodegradable materials. In *Proceedings of the 23rd International Symposium on Wearable Computers* (pp. 306–311). Association for Computing Machinery. https://doi.org/10.1145/3341163.3346938

Vermeer, C. M., Rossi, E., Tamis, J., Jonkers, H. M., & Kleerebezem, R. (2021). From waste to self-healing concrete: A proof-of-concept of a new application for polyhydroxyalkanoate. *Resources, Conservation and Recycling*, *164*, 105206. https://doi.org/10.1016/j.resconrec.2020.105206

Veron, A. C. G., Ibarria, M. M., & Lòpez, E. G. (2013). Bioluminescent algae and possible implications in architectural design. In PLEA 2013 – 29th conference: Sustainable architecture for a renewable future, Munich, Germany, 10–12 September 2013.

Yim, S. S., McBee, R. M., Song, A. M., Huang, Y., Sheth, R. U., & Wang, H. H. (2021). Robust direct digital-to-biological data storage in living cells. *Nature Chemical Biology*, *17*(3), Article 3. https://doi.org/10.1038/s41589-020-00711-4

Yu, Z. (2023). *A design framework for dermatitis-oriented bacterial cellulose textile in the fashion field* [Master thesis, Politecnico di Milano]. https://www.politesi.polimi.it/handle/10589/201315

Yu, Z., Rognoli, V., & Bertola, P. (in press). Biodesign in textile innovation: Emerging applications of bacterial cellulose to fashion. In A. Mateus

(Ed.), *What's Around Design? Strategic and speculative biodesign for a sustainable future: Volume one*. Springer.

Zhang, J., Hu, Y., Zhang, L., Zhou, J., & Lu, A. (2023). Transparent, ultra-stretching, tough, adhesive carboxyethyl chitin/polyacrylamide hydrogel toward high-performance soft electronics. *Nano-Micro Letters*, *15*(1), 8. https://doi.org/10.1007/s40820-022-00980-9

Zhang, Q., Xin, C., Shen, F., Gong, Y., Zi, Y., Guo, H., & Wang, Z. L. (2022a). Human body IoT systems based on the triboelectrification effect: Energy harvesting, sensing, interfacing and communication. *Energy & Environmental Science*, *15*(9), 3688–3721.

Zhang, X., Wang, X., Fan, W., Liu, Y., Wang, Q., & Weng, L. (2022b). Fabrication, property and application of calcium alginate fiber: A review. *Polymers*, *14*(15), 3227. https://doi.org/10.3390/polym14153227

Zhao, S., Guo, C., Kumarasena, A., Omenetto, F. G., & Kaplan, D. L. (2019). 3D printing of functional microalgal silk structures for environmental applications. *ACS Biomaterials Science & Engineering*, *5*(9), 4808–4816. https://doi.org/10.1021/acsbiomaterials.9b00554

Zhu, B., Wang, H., Leow, W. R., Cai, Y., Loh, X. J., Han, M.-Y., & Chen, X. (2016). Silk fibroin for flexible electronic devices. *Advanced Materials*, *28*(22), 4250–4265. https://doi.org/10.1002/adma.201504276

CHAPTER 3

DESIGNED INTERACTIVE LIVING SYSTEMS (DILS)

TRACING THE PATH FROM FOUNDATIONS TO REAL-WORLD APPLICATIONS

DOI: 10.4324/9781003366799-4

THE BIOLOGICAL IMPERATIVE: THE PATH TOWARDS DILS

In his essay "Our Biotech Future", published in the *New York Review* in 2007, theoretical physicist Freeman Dyson proposed that "The twentieth century was the century of physics, and the twenty-first will be the century of biology" (Dyson, 2007). Dyson envisioned biology as a transformative force capable of profoundly influencing various aspects of society, including the economy, ethics and human well-being. Writing just 5 years after the groundbreaking success of the Human Genome Project,[1] Dyson encouraged society to view biology as the next great "moonshot" for human progress. The beauty and complexity of biological systems have long served as a source of inspiration for designers, artists and scientists alike. For example, Antoni Gaudí's spiral and soaring structures draw directly from natural geometries such as helicoids and hyperbolic paraboloids, demonstrating how biology can shape architectural forms. Similarly, the Eastgate Centre in Zimbabwe mimics the self-cooling mechanisms of termite mounds, drastically reducing the energy required to ventilate the building. The lotus effect, observed in the water-repellent and self-cleaning properties of lotus leaves, has been used to create hydrophobic coatings for surfaces and textiles. In technology, gecko-inspired adhesives mimic the microscopic structures of gecko feet to provide strong, reusable, residue-free adhesion (Benyus, 2002; Kapsali, 2021). As Peters (2011) notes, biological systems provide models and drive innovative designs and theories that shape our understanding of and interaction with the world.

Classical biomimetics has traditionally focused on emulating nature, but in recent years, the field has evolved into two distinct approaches (Aziz & El Sherif, 2016). Top-down biomimetics, or design that looks to biology, starts with a human need or design problem and seeks solutions from biological systems (López Forniés & Berges Muro, 2012). In contrast, bottom-up biomimetics, or biology-influencing design, starts with specific biological features and translates them into human applications (Perricone et al., 2021; Lenau et al., 2018).

As biodesign takes center stage, it is clear that the future of biological integration in design and technology goes beyond simply imitating nature (Myers, 2012). Biodesign is the practice of integrating living organisms or biological systems into the design process, not just as inspiration but as active, functional components of the final product. It represents a deeper collaboration with nature, from taking inspiration from living systems to embedding them directly into objects and processes. This paradigm aims to replace traditional production methods with biological processes, with biofabrication emerging as the main driver (Pollini, 2021; Camere & Karana, 2018).

The increasing application of biological principles is paving the way for innovative paradigms in materials, production and design processes. This approach has led not only to the development of "living materials"[2] (Nguyen et al., 2018; Ginsberg & Chieza, 2018) and living artifacts (Karana et al., 2020; Zhou et al., 2022; Karana et al., 2023), but also to the expansion of the boundaries of interaction design (Gough et al., 2021). Recent work by Gough et al. (2020) further elaborates on the development of biodesigned systems within HCI, emphasising how living matter and computational interaction can converge to create the next generation of interactive, responsive artefacts. The integration of biology and technology in this field creates a fascinating fusion of the natural and the artificial, opening up new possibilities and experiences in human–computer interaction (HCI) Researchers have made significant progress in exploring how biological systems can be integrated into interaction design (Jeremijenko, 2012; Parkes & Dickie, 2013; Pataranutaporn et al., 2018; Ofer et al., 2021; Kim, 2021; Kim et al., 2021; Groutars et al., 2022; Zhou et al., 2022; Kim et al., 2023). Their work focuses on how biological systems' composition, mechanisms and metabolic processes can inspire innovative interaction design models. These contributions advance the field by developing cutting-edge interfaces and technologies that fuse biological and computational elements. This dynamic fusion blurs the distinction between living and non-living matter, creating a continuum of computational interactions that redefine the traditional boundaries between biology and technology. Ultimately, these innovations have the potential to transform the built environment by fostering new relationships between people, technology and living systems.

The Rise and Evolution of BioHCI

Recent advances in human-computer interaction (HCI) highlight the emergence of bioHCI, a field that integrates biology and technology to improve user interfaces and interaction experiences. This paradigm shift involves the use of biological processes and feedback mechanisms to create innovative, adaptive and sustainable systems. Foundational concepts such as "Tangible Bits" and "Radical Atoms", pioneered by Hiroshi Ishii,[3] have laid the groundwork for this evolution.

Tangible Bits aimed to materialize digital information and computational processes in physical forms but faced challenges due to the rigidity of physical materials compared to the adaptability of digital bits (Ishii & Ullmer, 1997; Ishii, 2008). To address these limitations, Ishii introduced Radical Atoms, envisioning materials that can be computationally reconfigured in real time (Ishii et al., 2012; Lakatos & Ishii, 2012). This concept proposed a future where all digital information could manifest physically, enabling direct interaction while opening pathways to programmable materials capable of integrating with biological systems. Radical Atoms represents a vision for the future of human–material interaction in which all digital information has a physical manifestation that allows direct interaction with it. This concept suggests the exploration of an alternative material reality that complements existing interaction models while opening the door to programmable materials that can converge with biological code.

Building on these ideas, Parkes and Dickie's (2013) concept of the "biological imperative for interaction design" further redefined the boundaries of HCI. They reimagined living and non-living matter as a continuum for computational interaction and recognized biological alternatives as opportunities to design, sense and display using natural mechanisms. This framework emphasized the creation of devices that were independent of traditional energy systems and manufacturing processes that were aligned with the planet's natural waste and renewal cycles, providing the basis for circular and sustainable production models. The recent development of the bioHCI framework, introduced by Pataranutaporn et al. (2018), extends these concepts by integrating biological materials into human–computer interfaces. This framework focuses on three key components: biological materials, intermediate platforms and user interaction. It reimagines biological materials as elements enabling the development of biodigital interfaces that bridge the gap between living systems and digital technology. Inspired by Ishii's Tangible Bits, Pataranutaporn and colleagues introduced the concept of "Living Bits" (Pataranutaporn et al., 2020), which incorporates microorganisms into HCI, marking a significant step towards unifying the biological and digital domains.

Their vision introduced the foundation for Designed Interactive Living Systems (DILS), which frame interactive systems based on living organisms, enabling a harmonious fusion of design, biology and computation.

These developments challenge the status quo of computation and materiality and offer opportunities for more sustainable and regenerative interactive components. In a world where digital technology is integral to modern society, DILS and the bioHCI framework highlight the potential to rethink interaction design through the lens of biology. By combining a design approach with biological and computational elements, they push the boundaries of what is possible at the intersection of these three domains, opening the way to a more sustainable and innovative future.

The Rise of Synthetic Biology and the DIYers Who Wanted to Play It

Suppose significant advances in tangible media and distributed manufacturing have been driven by the emergence of accessible digital fabrication technologies and the Maker Movement (Gershenfeld, 2005). In that case, it can similarly be argued that contemporary biodesign and living interaction design are supported by the foundational contributions of synthetic biology and the efforts of dedicated enthusiasts and community-driven initiatives seeking to apply this pioneering field of biology to DIY practices.

Synthetic biology, an interdisciplinary field within bioengineering, involves deconstructing and reassembling cells and biological processes to create novel systems capable of performing specific, useful functions (Garner, 2021). This modular and design-oriented approach to contemporary biology challenges the conventional image of researchers confined to microscopes and enables a wide range of applications, including the production of biomaterials, biofuels, biosensors and advanced biocomputing systems. Since its emergence in the early 2000s with its first practical applications, synthetic biology has played a pivotal role not only in facilitating the rise of movements such as DIYBio (Kuznetsov et al., 2012; Delgado, 2013) but also in enriching the field of bioHCI (human–computer interaction applied to biology) and significantly advancing the discourse around DILS. Similar to DILS, synthetic biology is characterized by a strong design-oriented approach that integrates biology, engineering, computer science and chemistry to create new biological systems or re-engineer existing ones. Its aim goes beyond understanding biological systems to transforming them into functional tools through rational design and innovative technologies. This seamless fusion of science and engineering is ushering in a new era of personalized and sustainable biological solutions, expanding the potential for interactions between the biological and digital domains and enriching the design vocabulary of DILS.

The proliferation of synthetic biology has catalyzed the formation of diverse communities of practitioners, including academics, industrial scientists, independent biohackers and artists.[4] The foundation for these advances was laid by the standardization of biological parts, (Rebatchouk et al.,1996; Arkin & Endy, 1999; Endy, 2005). These efforts led to the development of tools such as Biobricks,[5] OpenWetWare,[6] OpenPCR[7] and BigBio,[8] which enable modularity and scalability. At the same time, community-driven initiatives such as the IGEM Part Standard Registry – which now includes over 20,000 standardized biological components – illustrate how collaborative efforts can drive innovation and accelerate progress in the field (Smolke, 2009).

The democratization of synthetic biology has been enabled by the open sharing of physical biomaterials, online access to protocols and a collective ethos of inclusivity. This approach has fostered a more open and accessible community of practice than traditional scientific fields, breaking down barriers between academic science and independent experimentation. From Sophia Rooth's exploratory narratives in *Synthetic* (2017) to Susan Hockfield's technological analyses in *The Age of Living Machines* (2019), the history of synthetic biology highlights a dynamic interplay between technology and community, highlighting the central role of collaboration and grassroots innovation.

Alongside the early enthusiasm for initiatives such as the International Genetically Engineered Machine (IGEM) and the Personal Genome Project, the movement to democratize biology has encouraged experimentation and tinkering outside of professional settings, drawing inspiration from other DIY movements. This growing community of professionals and amateurs operates independently of institutional boundaries, promoting a hands-on approach that combines open-source ethics with the joy of exploring biological matter. In doing so, it challenges and redefines conventional practices in biology (Kuznetsov et al., 2012; Delgado, 2013). A critical component of this landscape is the creation of dedicated spaces for synthetic biology experimentation. Research by Gomez-Marquez and Hamad-Schifferli on distributed biological foundries for global health has highlighted the importance of community-driven spaces dedicated to life sciences, known as biospaces (Gomez-Marquez & Hamad-Schifferli, 2019). Spaces such as Biocurious, Genspace and Biotech Without Borders aim to promote scientific literacy and education by providing inclusive access to laboratories, tools and resources. Inspired by makerspaces, biospaces have been specifically adapted to the needs of biohacking, offering an operational model that emphasizes accessibility, safety and innovation.

Open to people with varying levels of expertise, these spaces provide training in basic microbiological techniques and essential laboratory skills, enabling participants to carry out biological experiments. Operating outside of traditional academic frameworks, biospaces are characterized by minimal infrastructure requirements and are supported by dynamic, engaged communities. This model reduces entry barriers and fosters an environment conducive to learning, experimentation and innovation, making life sciences more participatory and accessible.

Chapter 5 presents examples of biolabs and biospaces that embody the exploratory spirit of the human–computer interaction (HCI) community. This community is characterized by a constant search for new methods, models and aesthetics of interaction, driven by the convergence of biology, design and technology. Integrating biodesign with the democratization of scientific knowledge and skills has led to the emergence of BioHCI. This nascent field is redefining how humans interact with living systems and biological materials.

By promoting accessibility and standardization, synthetic biology has become a cornerstone of biodesign and living interaction design, providing tools, methodologies and design perspectives that align with the innovative DILS framework. These links emphasize shared values such as innovation, sustainability and interdisciplinary collaboration and position synthetic biology as a driving force in the evolution of bioHCI and DILS. This integration expands the potential for interaction between the biological and digital domains, opening new horizons for contemporary design.

PIONEERING DILS: A COLLECTION OF CASE STUDIES

DILS represent a groundbreaking frontier in design, where living systems are not only integrated into interactive systems, but also actively shaped through interdisciplinary collaboration. They often emerge from scientific and technological research and place designers at the center of the process. This approach enables them to experiment, innovate and create in interdisciplinary contexts, bridging biology, computation and design to ensure that these systems are scientifically viable, socially relevant and responsive to human needs and broader society goals.

Through DILS, design goes beyond traditional biological inspiration by embedding living organisms, biofabricated materials and biological processes directly into designed systems. These systems emphasize the interaction between users and their environment, placing living systems at the center of their function and behavior. Designers act as facilitators, translating biological principles into viable solutions that extend the boundaries of design to include sustainability, regeneration and interaction.

This chapter presents a collection of case studies that define and demonstrate the concept of DILS and show how design, particularly interaction design, is playing an active and pioneering role in its development. By analyzing over 40 case studies, these examples illustrate how the principles of DILS are materializing in real-world applications, merging design, biology, technology and interaction to drive innovation and sustainability. These projects challenge traditional distinctions between the natural and the artificial, offering solutions that are adaptive, circular and inherently ecological.

The case studies are categorized into seven key application areas: Electronics and Sensors, Medicine, Green Energy and Air Pollution Remediation, Lighting Design, Fashion Design, Visual Communication, and Entertainment.

This categorization highlights how DILS is crossing traditional disciplinary boundaries, moving from niche experimental projects to scalable, potentially mainstream applications. Designers are at the heart of this transition, tasked with envisioning and demonstrating how DILS can be seamlessly integrated into everyday life, increasing their accessibility and social acceptance.

Despite its transformative potential, the development and implementation of DILS face significant challenges. Scalability remains a critical issue, with many current projects limited to experimental or speculative prototypes. Ethical concerns about the use of living organisms and the potential ecological impact of large-scale deployment also need to be carefully addressed. Designers have a crucial role to play in these discussions, framing the ethical discourse and illustrating the possibilities and long-term benefits of DILS.

In this context, DILS represent not only a convergence of disciplines, but also a model for the future of innovation. They challenge conventional paradigms by demonstrating how design can transcend aesthetics and functionality to engage with complex biological and computational systems. While much of the work in this field remains speculative, history shows that designers have consistently driven the adoption of emerging technologies, making them accessible, scalable and aligned with social values.

Ultimately, DILS promise to redefine interaction design by bridging the gap between living systems and human-centered innovation in a way that is sustainable, regenerative and in harmony with natural ecosystems – in a word, more than human HCI. By aligning with broader trends in sustainability and regenerative practices, DILS mark a pivotal shift in design, offering solutions that meet the challenges of today while envisioning the possibilities of tomorrow.

Electronics and Sensors

The development of DILS represents a breakthrough shift towards ecological innovation in electronics and sensing. This transition is characterized by the convergence of electronics, synthetic biology and electromicrobiology, enabling the creation of living electronics and sensors that harness the natural functionality of living organisms – from bacteria and algae to human cells – for applications in electronics (Atkinson et al., 2022). In these systems, biological entities such as biomolecules, cells or cell communities are directly integrated into electronic circuits and act as active electrical components (Dunn, 2020). By embedding biological elements into electronic devices, this innovative approach opens up new possibilities for the development of sustainable DILS.

Biohybrid devices incorporating living electronic components at various scales are gradually dissolving the boundaries between living systems and artificial electronics (Zhang et al., 2020). These

DILS exhibit remarkable properties, including the ability to adapt to environmental changes, generate electrical signals, process information and perform functions such as replication, self-healing and evolution. This seamless integration of biology and technology not only enhances the capabilities of electronic systems, but also paves the way for sustainable, adaptive and responsive technologies.

Researchers funded by the Office of Naval Research (ONR) and led by Professor Derek Lovley[9] at the University of Massachusetts Amherst have made significant advances in engineering a common soil bacterium, *Geobacter*. Bioengineering has enabled this bacterium to produce ultra-thin, highly conductive and renewable electrical wires, known as "*Geobacter* nanowires". These groundbreaking achievements have been extensively documented (Lovley & Walker, 2019; Lovley et al., 2011; Bond & Lovley, 2003).

Geobacter nanowires are characterized by their incredibly small size – thousands of times thinner than a human hair. Made from non-toxic proteins, these nanowires are environmentally friendly and can be sustainably produced using resources such as solar energy, carbon dioxide or plant waste. Not only do these synthetic nanowires outperform their natural counterparts in terms of conductivity and size, but they also hold transformative potential for the field of bioelectronics. The applications of these bionanowires are vast, ranging from sensors to transistors to capacitors. Their versatility spans areas such as environmental monitoring and wearable sensing, while also representing a significant step towards renewable, environmentally friendly energy sources. This innovation demonstrates the feasibility of using microorganisms to create efficient electrical components and underlines their alignment with sustainable and environmentally conscious practices.[10]

At the Université Paris Diderot and ITODYS[11] in Paris, the BIOelectronics and Smart Surfaces (BiOSS) team has been investigating the potential of microalgae – specifically chlorophytes, cyanobacteria and diatoms – as solutions to water pollution. Found in marine and freshwater environments, these microalgae have an innate ability to detect a range of pollutants, including herbicides and heavy metals (Brayner et al., 2011; Antonacci and Scognamiglio, 2019). These photosynthetic microorganisms are remarkably sensitive to even trace amounts of pollutants (Kergoat et al., 2012), acting as effective living sensors for monitoring water quality. This research highlights the critical role that microalgae could play in environmental protection and sustainability efforts.

Meanwhile, a team at MIT has developed a new "living material": a flexible and resilient hydrogel sheet enriched with genetically engineered cells. These cells, mainly *Escherichia coli*, have been programmed to emit light when they detect certain chemicals.[12] Key to this innovation is the work of MIT's synthetic biology group, which focuses on programming *E. coli* cells for specific functions, such as detecting viruses and toxins. To overcome the challenges of maintaining cell viability in synthetic environments, the team in Professor Zhao's Soft Active Materials Laboratory has developed a hydrogel that supports cell life by resisting breakage and leakage. This hydrogel consists of a mixture of polymers and water that can support a high water content without losing its integrity, even under stress. By incorporating genetically modified *E. coli* into the hydrogel, the team has created a material that glows in the presence of certain chemicals and has demonstrated its application through various prototypes, including a glove with living sensors at the fingertips that glow when exposed to chemicals (Yuk et al., 2019). The project continues to explore the adaptation of this technology for broad applications, from forensics to health diagnostics, with a focus on integrating different types of bacteria to detect environmental toxins or skin diseases.

Supported by the Office of Naval Research, the National Science Foundation and the National Institutes of Health, this research not only paves the way for living materials in wearable sensors and diagnostic tools but also introduces a model for the design of these innovative devices, potentially revolutionizing fields such as environmental monitoring and medical diagnostics.

Viruses and DNA are also being explored as innovative tools to improve electronic materials and devices. Viruses have been used to modify the properties of materials, enabling the creation of nanostructures or nanowires that enhance the efficiency of transistors and solar cells (Mao et al., 2009). Similarly, DNA's unique self-assembly and sequence recognition capabilities make it a powerful resource for the development of living electronics and sensors. These properties enable the construction of highly tailored sensors and electronic circuits, unlocking new technological possibilities (Cheng et al., 2023).

Complex organisms, including plants, are increasingly being explored as candidates for the development of DILS as living sensors. Harpreet Sareen[13] and colleagues at the MIT Media Lab have pioneered an innovative perspective in interaction design, called "Cyborg Botany".[14] This approach leverages the inherent sensory and responsive capabilities of plants, as documented in their work (Sareen et al., 2019). Two key projects exemplify this research: Phytoactuators as output devices and Planta Digitalis for inputs.

Phytoactuators act as output devices, using a mobile application to stimulate plant responses (see Figure 3.1). For example, electrodes attached to a Venus flytrap can trigger its leaves to close, allowing controlled interaction between the plant and its environment. In contrast, Planta Digitalis acts as an input system, allowing plants to receive and respond to environmental information (see Figures 3.2 and 3.3). This is achieved by cultivating a conductive "wire" within the plant that provides a pathway for sensing and responding to external stimuli.

This innovative approach expands the potential for plants to act as motion detectors or environmental monitors. Applications of cyborg

Figure 3.1 Phytoactuators (2017) - Bioactuation achieved through electrodes integrated into plant leaves, enabling controlled movement and interactive responsiveness.
Credits: Images courtesy of Dr. Harpreet Sareen, Parsons School of Design, New York.

botany range from home security systems to military uses, where the unique sensory capabilities of plants can be harnessed for a variety of purposes. This research highlights the transformative potential of integrating plants into interaction design, bridging the gap between DILS and technological applications.

The Botanicus Interacticus project[15] is an example of an innovative method of transforming living plants into highly interactive media without harming or disrupting the plants themselves. This

Figure 3.2 Planta Digitalis (2019) – Detail of the plant-antenna interface illustrating the living sensing setup, where in vivo antennas enable real-time environmental signal detection through plant tissues.

Credits: Images courtesy of Dr. Harpreet Sareen, Parsons School of Design, New York.

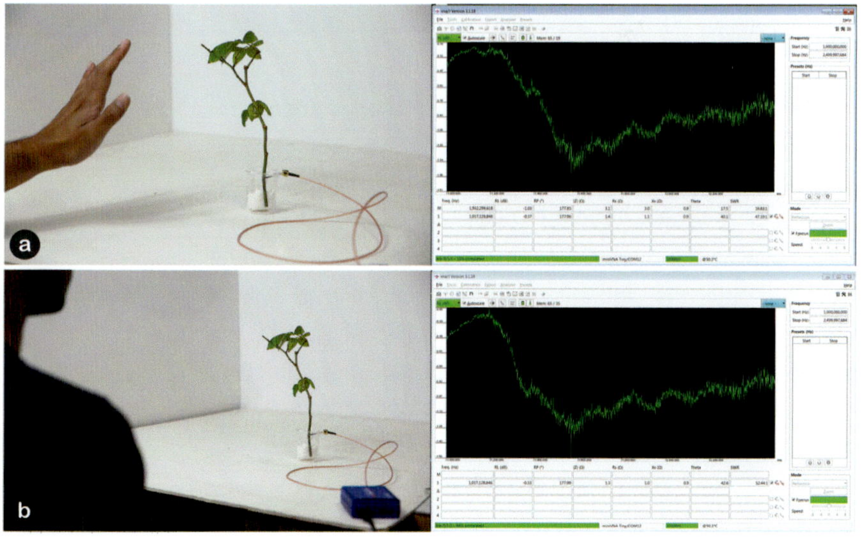

Figure 3.3 Planta Digitalis (2019) – Using plants as motion sensors through the integration of in vivo antennas that detect and transmit environmental stimuli.

Credits: Images courtesy of Dr. Harpreet Sareen, Parsons School of Design, New York.

non-invasive technology works by inserting a simple wire into the plant's soil, enabling a range of sophisticated interactions. For example, the biosensor can detect human proximity and the specific location of touch on the plant (Poupyrev et al., 2012).

Projects such as Botanicus Interacticus open up new avenues for the design of DILS and organic interfaces. By extending the possibilities for interaction with the natural world, this technology offers promising applications in fields such as art, education and interactive installations, enriching human engagement with living systems.

The integration of living organisms into security and detection systems further highlights their potential. In a groundbreaking study, researchers harnessed the olfactory capabilities of locusts, demonstrating their ability to detect and discriminate between explosive odors within milliseconds of exposure (Saha et al., 2020). This extraordinary sensory precision highlights the potential for using insect sensory systems in advanced detection technologies (to detect explosives, for example).

Similarly, researchers at the Central Institute of Fisheries Education in Mumbai made a significant breakthrough by developing a transgenic zebrafish for detecting heavy metals in aquatic environments (Pawar et al., 2016). This innovation involved the integration of a metal-responsive promoter with a fluorescent reporter gene into the zebrafish genome, enabling real-time monitoring of heavy metal contamination. This approach exemplifies the use of genetically modified organisms for environmental monitoring and demonstrates the versatility of living systems as biosensors.

These examples highlight the transformative potential of living systems in interactive and sensing technologies, bridging the gap between biology and technological innovation to address diverse societal challenges.

In a remarkable development, researchers funded by the Defense Advanced Research Projects Agency (DARPA) have developed a novel method for detecting hidden landmines using genetically modified *Escherichia coli* bacteria.[16] The project, led by scientists at the Hebrew University of Jerusalem, represents more than a decade of dedicated research, with significant progress (Belkin et al., 2017). The innovative technique uses bioengineered *E. coli* that emits light when exposed to by-products of trinitrotoluene (TNT), a key component of many landmines. In practical application, beads containing the modified bacteria are strategically distributed over a suspected landmine site. A laser is then used to detect the illuminated bacteria, accurately identifying the locations of concealed landmines (Cross, 2017). This method represents a breakthrough in landmine detection, combining biotechnology and advanced sensing techniques to provide a safer and more efficient alternative to traditional approaches. By harnessing the natural responsiveness of engineered microorganisms, this technique has the potential to increase the safety of demining operations while reducing the environmental and human risks associated with landmine clearance.

Living organisms are also providing innovative tools to enhance digital security. The stability and integrity of secure Internet connections rely heavily on cryptographic techniques that require the generation of random numbers. Random numbers fall into two categories: true random numbers, which are inherently unpredictable, and pseudo-random numbers, which are generated by deterministic processes. True random numbers, derived from physical phenomena such as white noise or radioactive decay, are essential for robust online security systems.

The Bacteria as Digital Mediator[17] project, led by Akvile Terminaite,[18] presents a groundbreaking approach to generating truly random numbers (see Figures 3.4–3.6). The method uses bacteria commonly found on the human hand. Using a handprint as a

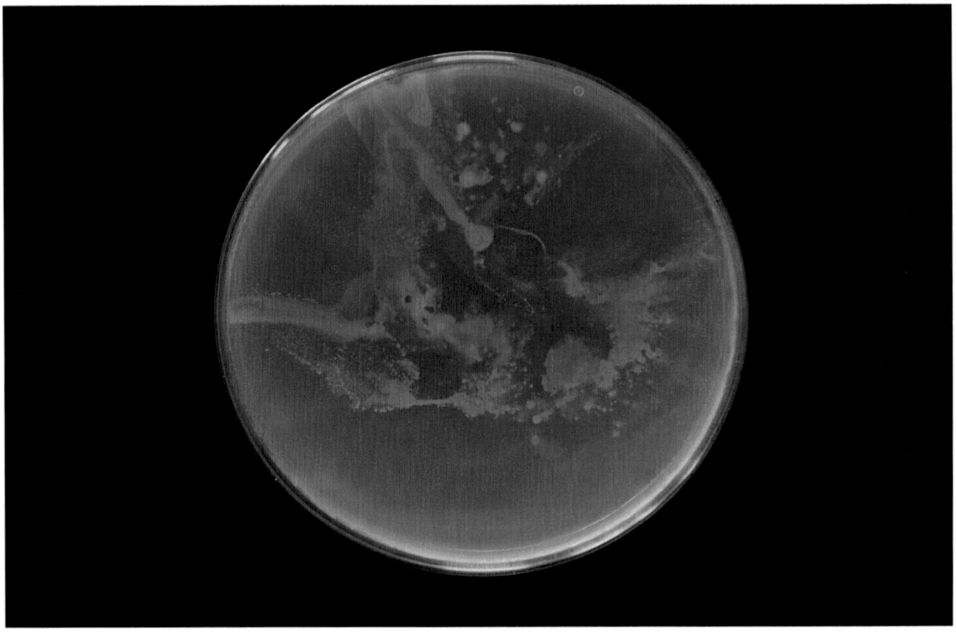

Figure 3.4 Bacteria as a Digital Mediator – Cultivation of hand bacteria for true random number generation. This image shows the growth of bacterial colonies used as the biological basis for generating true random numbers.
Credits: Akvile Terminaite, Experience Designer, RCA 2020, Royal College of Art, in collaboration with Open Cell, Erik Lintunen, and Phillip Quiza.

biological trigger, the project generates a sequence of truly random numbers. The inherent unpredictability of bacterial responses underpins this approach, providing a novel, biology-based solution to enhance cryptographic security.

This innovative use of living organisms marks a significant advancement in the field of cryptography. By exploring biological mechanisms as alternatives to conventional digital methods, the project opens new frontiers for secure communication, combining the randomness of natural processes with technological applications to strengthen digital security frameworks.

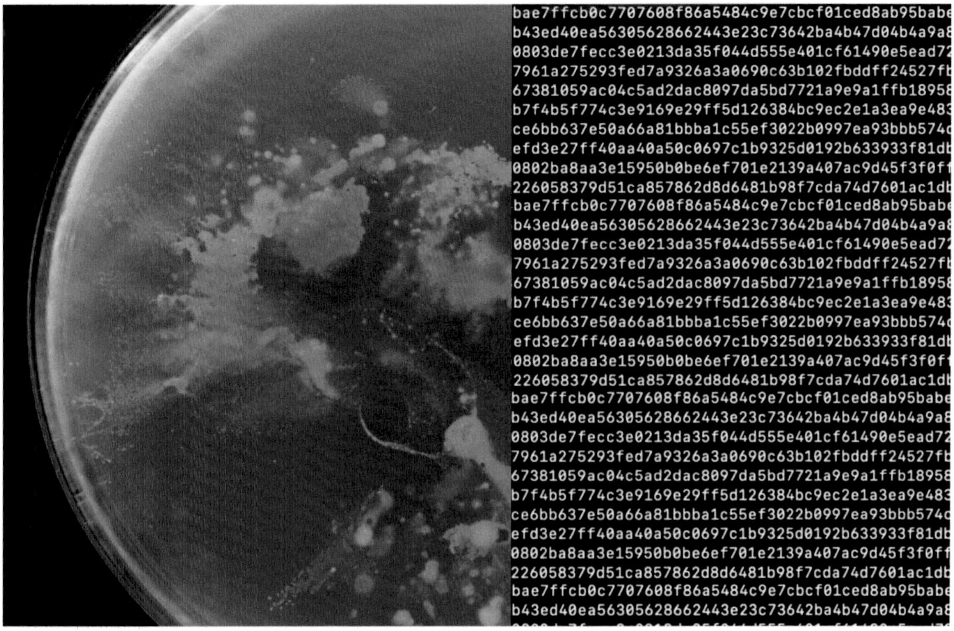

Figure 3.5 Bacteria as a Digital Mediator – Conversion of bacterial patterns into digital output. This image illustrates the process of translating bacterial growth patterns into sequences of random numbers for cryptographic applications.

Credits: Akvile Terminaite, Experience Designer, RCA 2020, Royal College of Art, in collaboration with Open Cell, Erik Lintunen, and Phillip Quiza.

Medicine

The integration of Designed Interactive Living Systems (DILS) into medical applications is transforming healthcare by exploring the dynamic interplay between living organisms, materials and technology to create innovative solutions. These systems are used in a wide range of applications, including diagnostic tools, therapeutic devices, smart implants and tissue engineering, exploiting the natural properties of cells and living organisms (Dong et al., 2023; Shang et al., 2021). By improving the diagnosis and prevention of disease, DILS offer a proactive approach to patient care, while expanding the possibilities for self-monitoring

Figure 3.6 Bacteria as a Digital Mediator (2020) – A close up view of the capsule design that holds the petri dish, collecting hand microbiome from the users.
Credits: Akvile Terminaite, Experience Designer, RCA 2020, Royal College of Art, in collaboration with Open Cell, Erik Lintunen, and Phillip Quiza.

and advanced diagnostics (Jiménez-Rodríguez et al., 2022; Goers et al., 2013).

One of the most promising roles for DILS in medicine is in therapeutic innovation. Engineered cells integrated into these systems can be programmed to deliver drugs on demand, reducing side effects and improving treatment outcomes. This capability enables highly personalized treatment regimens that respond to the specific needs of individual patients.

In addition to their diagnostic and therapeutic benefits, DILS offer potential solutions to medical sustainability challenges.

The healthcare sector has a significant environmental impact due to the extensive use of single-use plastics, medical devices and chemical waste (Eckelman & Sherman, 2016; Steenmeijer et al., 2022). Incorporating living organisms, together with biofabricated and biobased materials into medical practices can help address these issues by promoting more sustainable, waste-reducing solutions, ultimately reducing the environmental footprint of healthcare systems.

A notable example of DILS in medicine is the development of ingestible sensors using genetically modified bacteria. Researchers at MIT have developed these sensors to diagnose gastrointestinal conditions such as stomach bleeding, a common indicator of peptic ulcers. Traditionally, diagnosing internal bleeding requires invasive endoscopic procedures that are often costly, complex and require patient sedation.

The bacteria-on-a-chip[19] innovation integrates sensors from living cells with ultra-low-power electronic components, creating a DILS capable of translating bacterial responses into wireless signals. These signals can be easily monitored in near real-time using a smartphone (Mimee et al., 2018). Timothy Lu, an associate professor at MIT specializing in electrical, computer and biological engineering, highlights the significance of this breakthrough: "by combining engineered biological sensors with low-power wireless electronics, we can detect biological signals in the body and in near real-time, enabling new diagnostic capabilities for human health applications".[20]

This groundbreaking approach has the potential to revolutionize diagnostics by making procedures less invasive, more accessible and more patient-friendly. The MIT research team has further extended the capabilities of these sensors to detect specific blood components, such as hemoglobin, as well as markers of inflammation. These innovations have been validated through successful testing in pig models, demonstrating the significant promise of DILS in advancing medical diagnostics.

Bringing together biology, design, engineering and technology, DILS represent a transformative approach to medicine, offering more sustainable, personalized and efficient healthcare solutions. Their potential to redefine diagnostic and therapeutic practices places them at the forefront of future medical innovation.

One of the earliest and most well-known examples of DILS is the E.chromi project, a groundbreaking initiative that combines design and science within the emerging field of synthetic biology to create a living sensor network. In 2009, a team of seven undergraduate students at the University of Cambridge started a project to genetically engineer *Escherichia coli* (E. coli) bacteria to emit a spectrum of visible colors. Using standardized DNA sequences known as BioBricks, they engineered the bacteria to display colors such as red, yellow, green, blue, brown and purple. These color changes were designed to act as indicators of environmental conditions, with the bacteria turning red in the presence of toxins, for example, to signal water contamination. This innovative work won E.chromi the grand prize at the 2009 International Genetically Engineered Machine (iGEM) competition.[21]

Working with the scientific team, designers Alexandra Daisy Ginsberg and James King explored the potential future applications and ethical implications of E.chromi technology.[22] Through speculative design, they imagined a series of scenarios that demonstrate the impact of living sensors on areas as diverse as food additives, intellectual property, personalized medicine, security and environmental challenges.

One particularly innovative concept they proposed is the Scatalog, a low-cost system for personalized disease monitoring. This concept involves the ingestion of engineered bacteria via yoghurt, which colonize the gut and continuously monitor for the presence of specific chemical markers associated with various diseases. When a marker is detected, the bacteria produce a visible alarm signal by changing the color of the individual's feces. Originally developed as a speculative tool to stimulate critical discussion within the synthetic biology community, the Scatalog has since inspired research into engineered probiotics, bringing this groundbreaking approach to disease diagnostics closer to reality.

These speculative narratives play a critical role in evaluating the societal implications of E.chromi, providing a framework for considering how such technologies might shape everyday experiences and social norms. By addressing the ethical, practical and cultural dimensions of E.chromi, Ginsberg and King highlight the broader significance of DILS in tackling complex societal challenges.

The collaboration behind E.chromi underscores its role as a DILS, demonstrating how such systems can integrate thoughtful design at both the genetic and societal levels. This project exemplifies the potential of interdisciplinary collaboration between artists, designers and scientists, and sets a precedent for the development of innovative technologies that balance scientific progress with critical societal reflection. E.chromi serves as a compelling model for future projects at the intersection of synthetic biology and design, demonstrating the transformative possibilities of such synergies.

Building on the innovative groundwork laid by projects in this expanding field, Giulia Tommasello's Future Flora project[23] aims to promote a symbiotic relationship between humans and beneficial microorganisms. Specifically, the project addresses the treatment and prevention of vaginal infections such as Candida vulvovaginitis (CVV), a condition that affects approximately 75% of women at least once in their lifetime.

Figure 3.7 Future Flora (2016): A living culture pad designed to help women balance their vaginal flora, aiding in the prevention and treatment of Candida infections. The project was awarded the STARTS Prize 2018 by EU Horizon 2020 for artistic exploration and, since 2019, has been part of the permanent collection of the Design Department at the Museum of Contemporary Art (MAK), Vienna.
Designer and Creator: Giulia Tomasello. External Expert: Biologist Arian Mirzarafie-Ahi. Graduate project: Material Futures Master's Course, Central Saint Martins College of Art & Design.
Credits: Giulia Tomasello.

Future Flora introduces a do-it-yourself (DIY) kit that allows individuals to cultivate bacterial pads enriched with strains of Lactobacillus bacteria known for their ability to inhibit the proliferation of Candida albicans (Figure 3.7). When applied to the female genital area, these pads promote the restoration of beneficial microflora by creating an environment inhospitable to pathogenic organisms. This replenishment of the vaginal epithelial microflora not only helps to treat infections but also prevents their recurrence, providing a sustainable and proactive approach to personal health care.

This initiative represents a pioneering convergence of biology and health technology, empowering women to engage in an intimate, informed and proactive form of self-care. By facilitating interaction with one's own body through the cultivation and interpretation of feedback using DILS, Future Flora demonstrates how innovative biotechnological solutions can enhance personal health and well-being (Figure 3.8).

Future Flora also highlights the potential of cultivating living organisms in the home environment as a pathway to improved health. By encouraging individuals to engage with their own biology, the project provides a model for proactive and informed self-care, advancing the role of DILS in creating accessible, sustainable and personalized healthcare solutions. This case study demonstrates how the integration of biology and design can transform traditional health practices into empowering, user-driven experiences.

Green Energy and Air Pollution Remediation

Energy is a fundamental necessity in modern society, playing a pivotal role in driving economic and social progress while significantly improving the quality of life (Bilgen, 2014). However, the predominant reliance on fossil fuels in recent decades has led to severe environmental degradation, including air pollution, deforestation, acid rain, ozone depletion and radioactive emissions (Dincer, 1999; Shamoon et al., 2022). This dependence not only contributes to climate change, but also poses significant risks to human health, particularly through air pollution in urban and residential areas.

The urgent need for sustainable energy solutions is underlined by the principles established by the Brundtland Commission[24] (1987), which emphasize the importance of meeting current energy needs without compromising the ability of future generations to meet

Figure 3.8 Future Flora (2016): Designed for women embracing control over their bodies, Future Flora promotes a new, intimate approach to self-care by encouraging active participation in the culture and understanding of science. The project envisions clothing and accessories as living ecosystems that support and balance the skin's microflora.

Credits: Tom Mannion.

their own needs. Fossil fuels, which are finite and environmentally destructive, require an urgent transition to renewable energy sources that minimize pollution and reduce carbon emissions. This transition is not only a response to the depletion of finite resources, but also a critical strategy for addressing the environmental crises exacerbated by decades of reliance on "dirty" energy sources.

Energy sources are typically categorized as "dirty" and non-renewable, such as coal and natural gas, or "clean" and renewable, including geothermal, wind and solar energy.[25] While renewable energy alternatives are widely regarded as more environmentally friendly, it is important to recognize that they too have environmental impacts. For example, the infrastructure required to produce wind and solar energy can disrupt ecosystems and consume significant resources. This highlights the complexity of the transition to sustainable energy systems and the need for a nuanced understanding of the environmental trade-offs associated with all forms of energy production (Jaiswal et al., 2022).

Designers and researchers are increasingly addressing these challenges by developing holistic solutions inspired by natural processes. Biomimicry and other nature-based approaches have gained prominence for their ability to emulate the efficiency and sustainability inherent in ecosystems. This innovation-driven, eco-conscious mindset is key to developing energy systems that are not only cleaner, but also more harmonious with the environment. These efforts represent a shift towards integrating technology with natural systems to reduce humanity's ecological footprint while meeting energy needs in a sustainable manner.

Nature-inspired strategies have the potential to transform energy production in line with broader sustainable development goals. This approach highlights the possibility of synergy between technology and nature, paving the way for energy systems that are both effective and environmentally responsible. By combining innovative design, scientific research and ecological principles, the search

Figure 3.9 Geomerce (2015): Cultivating the Future of Sustainable Mining. The project by Gionata Gatto and Giovanni Innella leverages hyperaccumulator plants for environmentally sustainable metal extraction and soil remediation, transforming agricultural practices into a model for circular resource recovery.
Credits: Matteo Cremonini.

for sustainable energy solutions offers a promising way to address the pressing global energy challenges of the twenty-first century.

The Geomerce project,[26] developed by Gionata Gatto and Giovanni Innella (Figure 3.9), explores the innovative concept of phyto-mining, exploiting the extraordinary capabilities of hyperaccumulator plants (Brooks & Robinson, 1998). These plants have the unique ability to absorb and concentrate heavy metals such as zinc, copper and nickel through their root systems. Once absorbed, these metals can be recovered by processing the harvested plant biomass, allowing them to be reused in industrial applications. This dual-purpose approach not only assists in the remediation of soils contaminated with heavy metals but also facilitates the sustainable recovery of valuable resources.

By rethinking traditional agricultural practices, Geomerce positions farmland and crops as economic assets by assigning a financial value to the metals sequestered in plant biomass. This innovative fusion of agriculture and economics introduces a paradigm shift in which farming decisions are influenced not only by traditional agricultural metrics, but also by fluctuations in global financial markets and advances in scientific research. This approach highlights the economic potential of integrating environmental sustainability with resource recovery. It also highlights the remarkable interactive capabilities of plants, in particular their natural ability to identify and exploit specific environmental niches. By harnessing these intrinsic biological functions, Geomerce exemplifies the potential of DILS, where living organisms are actively involved in achieving sustainable and regenerative outcomes. By strategically integrating plant biology with economic modeling, Geomerce opens up new avenues for developing extraction methods that are both efficient and environmentally responsible. This synergy between biological systems and economic frameworks not only offers groundbreaking solutions to the dual challenges of environmental remediation and resource recovery but also demonstrates how DILS can redefine our relationship with natural systems by embedding living organisms as active, functional components in innovative design processes.

The Institute for Advanced Architecture of Catalonia (IAAC)[27] has introduced a groundbreaking prototype photovoltaic cell that uses algae as a sustainable alternative to traditional solar panel materials. The BioCatalytic Cell project[28] is spearheading the research and application of innovative, sustainable technologies for solar energy capture and electricity generation (see Figure 3.10). At the heart of the project is the use of photosynthetic microorganisms, such as algae, as viable means of harnessing solar energy.

This initiative integrates a biophotovoltaic film system with an energy-sensing algal cell housed in a panel designed to rotate, ensuring uniform algal growth and optimal exposure to sunlight.

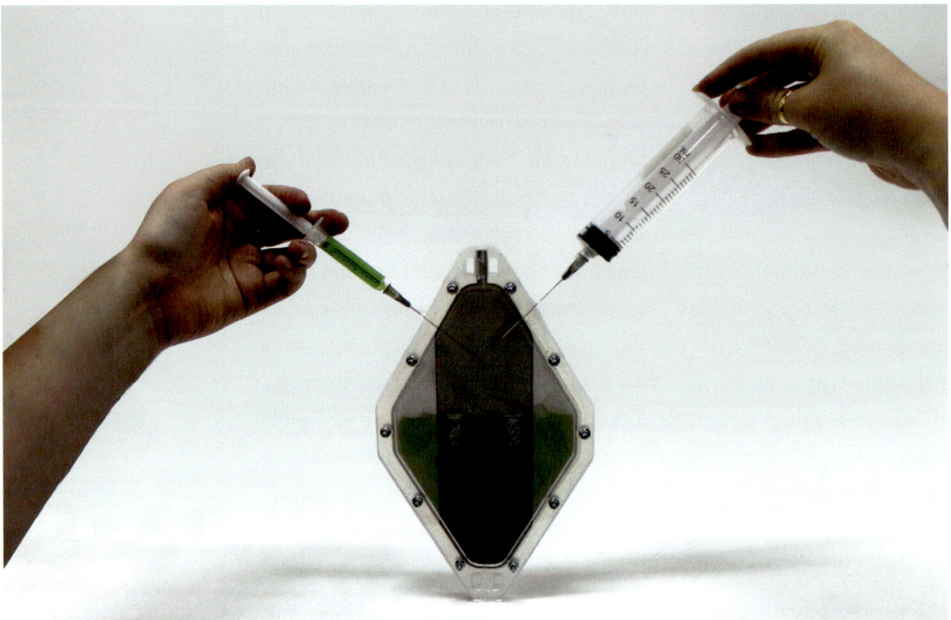

Figure 3.10 **BioCatalytic Cell (2017): A Designed Interactive Living System that employs a biophotovoltaic film to create an adaptive, energy-producing algae cell. The project harnesses the photosynthetic ability of microorganisms to capture solar energy and generate electricity.**
Designers: **Thora H. Arnardottir, Jessica Dias, Christopher Wong. Advisors: Chiara Farinea, Paolo Bombelli, Carmelo Zappulla, Maria Kupstova. Biology Support: Núria Conde Pueyo. Guest Advisor: Claudia Pasquero.**
Credits: Thora H. Arnardottir.

Transparent cell panels incorporate metal and fiber meshes to efficiently capture and transport electrons produced during photosynthesis. To support the viability of the algae, the panels act as sealed containers with small vents and carbon filters that allow oxygen exchange to maintain an ideal growth environment. In addition, a system of cables is included to collect the electricity generated by the clusters of algae panels.

The primary objective of the BioCatalytic Cell project is to power the system itself and to collect extensive data on algal energy production under a variety of conditions. This research will provide critical insights into the capabilities of photosynthetic microorganisms for energy production. The project represents a significant step towards the development of fully autonomous, living, interactive systems (DILS) for sustainable solar energy capture and use. By exploring organic alternatives to traditional, non-renewable solar panel materials, the BioCatalytic Cell project aligns with the broader goals of sustainability and regenerative practices outlined in previous research (McCormick et al., 2011).

Furthermore, as a designed interactive living system, the BioCatalytic Cell exemplifies the seamless integration of living organisms into technological frameworks. This synergy underscores the potential for biologically driven solutions to address critical energy sustainability challenges while advancing the use of renewable, interactive and environmentally responsible systems.

In a groundbreaking scientific development, researchers from Imperial College London, the University of Cambridge and Central Saint Martins have pioneered the use of cyanobacteria as a novel ink for printing on carbon nanotubes. This innovation introduces the concept of bio-solar panels[29] that can be printed onto wallpaper substrates, representing a significant advance in sustainable energy production and design. Cyanobacteria, ancient microorganisms renowned for their photosynthetic abilities, have historically played a key role in creating the Earth's oxygen-rich atmosphere over billions of years. The UK research team has demonstrated the successful use of cyanobacteria as an ink, enabling precise pattern printing on electrically conductive carbon nanotubes using advanced inkjet technology. Remarkably, the cyanobacteria maintain their photosynthetic activity after the printing process, enabling the generation of electrical energy over an extended period of time, with power output sustained for up to 100 hours (Sawa et al., 2017).

This prototype bio-solar panel, about the size of an iPad, is capable of powering low-power electronic devices such as digital watches or small LED lights. The project highlights the potential of integrating biological systems with cutting-edge technology to create energy solutions that are both sustainable and adaptable. This achievement not only highlights the synergy between biological processes and technological innovation but also sets the stage for future exploration in the field of DILS.

Recent advances in bioelectricity generation have seen algae and bacteria – long celebrated for their ability to produce electricity – joined by innovative competitors in the field. A groundbreaking study published in *ACS Applied Materials and Interfaces* (Shlosberg et al., 2022) by researchers from the Schulich Faculty of Chemistry at the Technion-Israel Institute of Technology highlights the untapped potential of plants, particularly water-rich succulents, as novel sources for bioelectricity production.[30] The research focused on the succulent plant *Corpuscularia lehmannii*, which transforms its leaves into living solar cells. By using the plant's natural reservoirs of water and nutrients as both a food source and electrolyte solution, the team mirrored the function of traditional electrochemical cells. This pioneering approach not only advances sustainable solar panel technologies but also reimagines how biological systems can contribute to environmentally friendly power generation.

Another innovative foray into bioelectricity is Electric Life,[31] a project by Teresa van Dongen[32] that exemplifies the integration of sustainable and bioderived energy sources. Powered by a "bacterial battery" developed in collaboration with the University of Ghent, the system captures electrons released during the metabolic processes of microorganisms to generate electricity. This bioluminescent system can power up to three light bulbs from a single battery unit, emphasizing a closed-loop and eco-conscious design (see Figures 3.11 and 3.12). Electric Life's sustainability relies on its users, who act as caretakers by adding nutrient-enriched tap water and a small amount of vinegar weekly to maintain the bacterial ecosystem. This interaction transforms maintenance into a meaningful engagement with a living energy source, embodying the principles of DILS (Ooms et al., 2022).

Figure 3.11 Electric Life (2019): Overview of the installation showing two self-standing light structures with microbial battery units at the base and luminous modules mounted on slender arms, illustrating the scale and spatial configuration.

Credits: René Gerritsen. Courtesy of Teresa van Dongen.

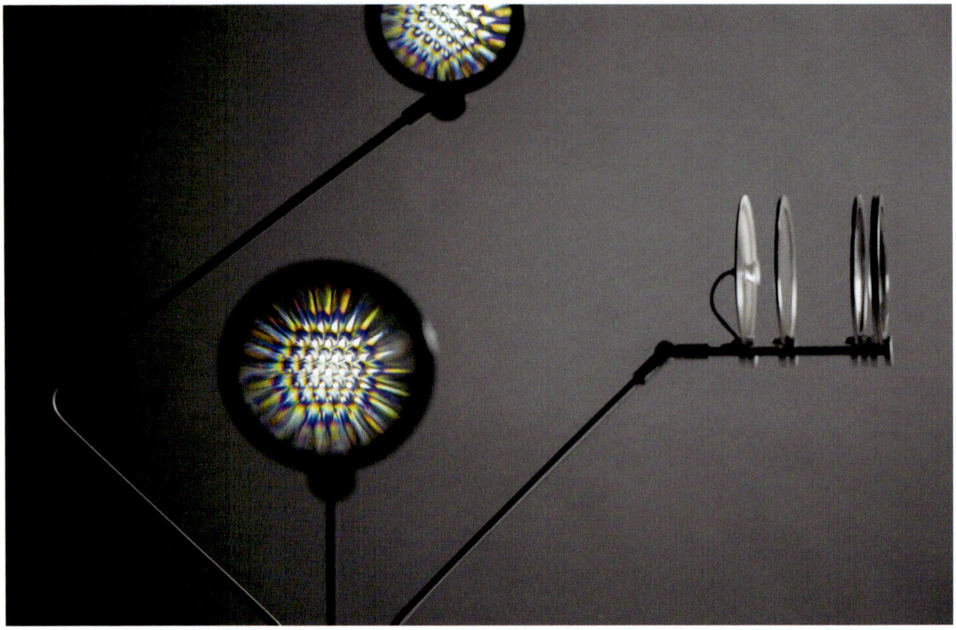

Figure 3.12 Electric Life (2019, detail): Close-up of the lighting modules and optical components, showing the distinctive radial pattern produced by the lenses and their connection to the bioelectric system.
Credits: René Gerritsen. Courtesy of Teresa van Dongen.

Biodesigner Teresa van Dongen has advanced the field of microbial energy by exploring the unique capabilities of bacteria living in the sedimentary depths of rivers and lakes. These specialized microorganisms emit electrons as a by-product of their metabolic processes, offering a promising avenue for the development of sustainable, domestic power generation. Van Dongen's research is an example of innovative approaches to green energy in line with the principles of sustainability and regeneration. Her work highlights the transformative potential of integrating biological systems into energy solutions, reinforcing the role of DILS in rethinking energy production and promoting environmentally responsible practices.

Another notable innovation is Living Light,[33] presented by Nova Innova in collaboration with the biotech designer Ermi van Oers at Dutch Design Week 2017. It is a revolutionary technology that uses microorganisms to convert the chemical energy from plant photosynthesis into electricity, operating independently of traditional power grids. Living Light is specifically designed to be self-sustaining and operating autonomously. At its core, Living Light consists of a plant in a glass enclosure. As the plant photosynthesizes, it releases organic compounds into a soil chamber where bacteria, powered by a microbial fuel cell, convert these compounds into electrons. These electrons travel through a wire to illuminate a ring of LEDs that light up when the plant's leaves are touched, creating an interactive experience (Figure 3.13).

This innovation demonstrates a concept of mutualistic care, suggesting a symbolic relationship between human interaction and the livingness of the artifact, as some scholars have noted (Karana

Figure 3.13 Living Light (2017): A self-sustaining lamp that generates electricity through plant photosynthesis, designed by Ermi van Oers.
Credits: Nova Innova.

et al., 2020). For example, the gentle touch of the leaves directly influences the intensity of the LED lights. Van Oers has the plan to extend this microbial energy technology to public spaces, starting with a collaboration with the city of Rotterdam to light a park. Applications of the technology could revolutionize urban and rural energy production, from streetlights powered by trees to rice fields in Indonesia producing both food and electricity.[34]

These projects highlight the transformative potential of harnessing natural processes for sustainable energy solutions. By designing DILS that integrate human interaction with ecological functionality, they create opportunities for a more symbiotic relationship between humanity and the environment.

Similar to the principles that underpin Living Light, another recent Nova Innova project, POND, represents a pioneering fusion of environmental science, design and technology, culminating in the world's first floating network powered by water's own microbes. This innovative project not only generates sustainable energy but also serves as a communicative bridge between aquatic ecosystems and humans, highlighting the intricate relationships within our environment. At the heart of POND lies microbial fuel cell technology. In natural water bodies, bacteria decompose organic matter, releasing electrons in the process. POND harnesses these electrons to generate electricity without disrupting the existing ecosystem. This harvested energy powers sensors that continuously monitor water quality parameters. The collected data is then translated into a captivating light display, with the colors and patterns reflecting the water's health status. For instance, a vibrant blue glow might indicate optimal conditions, while a shift to red could signal potential pollution or imbalance. Beyond its technological achievements, POND addresses pressing environmental concerns. Globally, unsafe water contributes to numerous health issues, and in Europe alone, 60% of water bodies are significantly polluted, posing risks to both ecosystems and public health. By illuminating the real-time health of water bodies,

POND fosters a deeper public awareness of water as a living entity that requires diligent care. This visual representation transforms abstract environmental data into an immediate, understandable and engaging experience for communities.

In recent years, urban environments have increasingly embraced green and biophilic philosophies to reconnect residents with nature, combat biodiversity loss and address urban pollution. Integrating trees into urban landscapes has emerged as a key strategy, particularly in the face of challenges such as urban expansion, which exacerbates the heat island effect and deteriorates air quality (Von Thaden et al., 2021). Trees act as natural air purifiers, producing oxygen through photosynthesis and acting as carbon sinks by absorbing atmospheric carbon dioxide, playing a vital role in mitigating global warming. Their contribution goes far beyond aesthetics, offering tangible solutions to some of the most pressing environmental challenges facing urban areas.

Building on this foundation, DILS propose an interdisciplinary approach that integrates natural systems such as trees and algae into urban design. DILS promote a collaborative synergy between biological processes and urban planning, fostering sustainable and livable urban environments. By exploiting the interactive capabilities of living systems, DILS not only improve environmental health, but also redefine the role of design in addressing climate resilience and ecological restoration.

The inclusion of algae-based technologies within this framework further illustrates the innovative potential of DILS. Algae, known for their ability to sequester carbon dioxide and produce oxygen, serve as natural air purifiers while also providing bioenergy solutions. Their integration into urban infrastructure – such as facades, public spaces and water systems – demonstrates how biological processes can be harnessed for environmental remediation. This holistic strategy is in line with the principles of sustainability and regeneration, promoting green infrastructure that beautifies urban landscapes while addressing critical issues such as air pollution, climate change and urban heat.

By embedding these living systems into the fabric of urban design, DILS provide a model for reimagining cities as dynamic ecosystems that actively interact with their environment. This approach not only increases urban resilience but also fosters a deeper connection between people and nature, creating healthier, more sustainable cities for future generations.

EcoLogic Studio,[35] a pioneering biodesign studio founded in London in 2005 by Claudia Pasquero and Marco Poletto, has been at the forefront of integrating DILS into urban landscapes, advancing the vision of a carbon-neutral city. Their innovative projects harness the unique ability of microalgae to absorb carbon dioxide and release oxygen, making a significant contribution to environmental sustainability while rethinking the role of living systems in urban design. A notable example of their work is the Habitat One installation[36] at the Hyundai MotorStudio in Seoul, also known as Tree One. This living, parametric structure is constructed from algae-based biopolymers and consists of 40 glass photobioreactors filled with cyanidium algae cultures. As a "living machine", Tree One[37] has the carbon sequestration capacity of 12 mature trees and demonstrates the transformative potential of microalgae in urban environments (see Figures 3.14 and 3.15). By integrating biological processes into architectural forms, this project exemplifies how DILS can redefine urban infrastructure to mitigate environmental challenges such as air pollution and climate change.

Another exemplary EcoLogic Studio project is H.O.R.T.U.S. XL,[38] which further illustrates the dynamic interaction between human activity and microorganisms. This installation, teeming with cyanobacteria, actively sequesters CO2 while releasing oxygen, offering visitors an immersive experience that demonstrates the seamless integration of biology and digital technology. Visitors

Figure 3.14 **Tree One (2023): A large-scale installation by EcoLogicStudio, composed of robotically 3D-printed algal biopolymers, lab-grade borosilicate glass, and living cultures of Cyanidium caldarium in liquid medium. Dimensions: 670 × 1,350 × 620 cm.**

Credits: Synthetic Landscape Lab, Innsbruck University.

Figure 3.15 Tree One (2023, detail) – Close-up view of the installation by EcoLogicStudio, showing the intricate network of robotically 3D-printed algal biopolymers and the fluid medium containing Cyanidium caldarium cultures.
Credits: Synthetic Landscape Lab, Innsbruck University.

are invited to engage with the system through the simple act of breathing, fostering a tangible connection between human presence and the functioning of these living systems.

These projects highlight EcoLogic Studio's commitment to exploring the synergy between natural systems, technological innovation and human interaction. By embedding DILS in urban environments, their work pushes the boundaries of sustainable design and presents a compelling model for cities as living ecosystems capable of dynamically interacting with their surroundings. This approach not only enhances ecological resilience but also fosters a deeper understanding of the interdependence between humanity and the natural world, reinforcing the critical role of DILS in shaping the cities of the future.

Another example of how we can integrate living systems into the cities of the future is Urban Reef, an innovative project that tackles urban air pollution and biodiversity loss by designing living, 3D-printed structures that transform cityscapes into functional ecosystems. These "reefs" use biobased materials and design algorithms that interact with the urban environment to restore air quality and promote sustainable microclimates. At the core of Urban Reef's approach is its ability to collect and distribute rainwater, creating self-sustaining microenvironments. These structures provide habitats for plants, mosses and microorganisms, which actively absorb carbon dioxide, release oxygen and filter particulate matter, contributing to cleaner urban air. The reefs also help regulate temperature and humidity in their immediate surroundings, counteracting the heat island effect commonly found in cities. Urban Reef's 3D printing technology allows for the customization of structures that passively interact with sunlight, water and nutrients, creating diverse spaces for life to flourish. By embedding ecological restoration into urban infrastructure, these reefs serve not only as air-purifying systems but also as green energy solutions. Their reliance on natural cycles, such as water flow and microbial interactions, requires no additional energy inputs, making them a sustainable and scalable solution for urban areas.

The Lichenized Biosystem, a project developed by Luz Wallace in 2023 at the Institute for Public Architecture on Governors Island, redefines urban architecture by transforming building facades into living, energy-generating ecosystems. This modular system integrates lichen-covered biomaterials with enzymatic biofuel cells, creating a self-sustaining structure that not only enhances biodiversity but also produces renewable energy. By embedding biofuel cells beneath the lichen layer, the system captures electrons generated during photosynthesis, converting them into electricity and offering a green energy alternative for urban infrastructure.

Lichen, a highly resilient symbiotic organism composed of mycelium and algae, is one of nature's most effective CO_2 sequestration agents. Despite covering only 8% of the Earth's surface, it captures nearly a quarter of global CO_2 emissions and releases oxygen, significantly improving air quality. Unlike traditional green facades that demand high maintenance and irrigation, lichens thrive on minimal resources, requiring no soil and adapting to harsh environments. Beyond carbon capture, the Lichenized Biosystem turns facades into carbon sinks, actively offsetting emissions from the built environment. The system is composed of biodegradable and recyclable biosubstrates, ensuring that it aligns with circular economy principles. The porous biomaterial structure supports lichen growth while fostering a symbiotic relationship between architecture, nature and humans. In polluted urban settings, an initial carbon-aerated layer can shield lichens until they stabilize and improve air quality. Designed for modularity and long-term adaptability, the system allows for removable and replaceable cultivation plates, ensuring sustainability and ease of maintenance. Even in extreme conditions, lichen enters dormancy rather than dying, resuming growth once favorable conditions return. This resilience makes it an ideal candidate for climate-adaptive building solutions (Wallace Luz, 2024).

In the case studies presented and others yet to be explored, the nature of user interaction with both objects and living organisms reflects a paradigm shift from conventional norms. Interaction is no longer limited to static engagement with objects but extends to dynamic interactions with the living organisms integrated into the design. Users engage with the outcomes generated by these DILS, such as improved air quality or energy production, illustrating a novel and functional mode of interaction between humans, objects and organisms.

Future scenarios envisage microorganisms not only populating building facades or inhabiting large-scale, air-purifying 3D-printed sculptures, but also becoming an integral part of the everyday human environment. This vision is exemplified by the Living Things project,[39] conceived by Ethan Frier and Jacob Douenias, which presents a pioneering domestic installation. Spirulina algae thrive in custom-designed glass bioreactors that are seamlessly integrated into household furniture. This innovative approach fosters a mutualistic and symbiotic relationship between humans and microorganisms, dissolving traditional boundaries between living spaces and the microscopic life forms that share them. Such integrations highlight the potential of DILS to redefine the relationship between biological and human-designed systems, embedding living processes into everyday life.

The convergence of biotechnology, design and engineering continues to shape the future of indoor environments. Julian Melchiorri's Exhale,[40] a bionic chandelier, is an example of this interdisciplinary synergy. This innovative chandelier incorporates leaf modules that house living microalgae that actively improve indoor air quality through photosynthesis. Beyond its aesthetic and lighting functions, Exhale stimulates the microalgae to absorb carbon dioxide and release oxygen, creating a symbiotic relationship between the object and its surroundings. Another example of how living organisms could enhance indoor well-being is the 2-week-long experiment conducted by Elvin Karana's team.

This study, called Living with Cyanobacteria, investigates the use of cyanobacterial living artifacts as an innovative approach to air purification. Designed to metabolize carbon dioxide and release oxygen through photosynthesis, these artifacts not only purify the air but also encourage new modes of ecological care in domestic spaces. The artifact, crafted with pliable, adhesive materials and color-changing properties, visually signals the microbes' well-being, inviting users to actively engage in its maintenance.

Participants in the study lived with the artifact for 2 weeks, adapting its placement to optimize light and air conditions. The artifact's materiality – a combination of silicone rubber and pH-sensitive jelly – proved central to fostering both curiosity and care. Observing color changes motivated participants to care for the microbes, integrating these tasks into their daily routines. This case demonstrates how living artifacts can spark ecological awareness and mutualistic care practices, even as their practical impact remains nascent. By balancing form, function and materiality, such designs offer new pathways for integrating sustainability into everyday life. Future iterations could enhance durability, functionality and interactivity to better align human and microbial needs.

After having analyzed how DILS could contribute to restore urban and indoor environments and how humans could benefit from it, in the last case study of this section, the Loop Living Cocoon,™ we want to show how also humans could contribute, with their death, to make environments and natural ecosystems healthier.

The Loop Living Cocoon,™ by Loop Biotech, represents a groundbreaking shift in funeral practices, directly addressing environmental sustainability and air pollution by integrating the natural lifecycle into burial traditions (see Figures 3.16 and 3.17). The Living Cocoon is made from mycelium, capable of decomposing organic material while simultaneously restoring the environment.

Figure 3.16 Loop Living Cocoon™ (2020). A biodegradable coffin designed by Loop Biotech to promote a sustainable, circular approach to human burial. Made from mycelium—the root structure of fungi—the cocoon accelerates the body's natural decomposition and enriches the surrounding soil with nutrients.

Credits: Loop Biotech.

Figure 3.17 Loop Living Cocoon™ (detail). Close-up of the mycelium-based structure and the moss layer inside the coffin, illustrating the living material's texture and its role in facilitating biological regeneration.
Credits: Loop Biotech.

Mycelium's ecological role is pivotal in this context. It enhances soil biodiversity and neutralizes toxins, creating healthier ecosystems. Traditional burial practices often involve the use of chemically treated wood and embalming fluids that can contaminate the earth and, in turn, harm surrounding environments. In contrast, the Living Cocoon decomposes in just 45 days, enriching the soil while removing pollutants. Additionally, the production process, using locally sourced organic materials and grown within days, reduces carbon emissions and dependence on resource-intensive materials like hardwoods.

As mycelium breaks down organic material, it captures nutrients and can even sequester harmful substances, potentially reducing localized air and soil pollution. This aligns directly with green energy principles by supporting regenerative cycles of life, rather than extracting from or damaging the earth. The Living Cocoon is a practical, symbolic and ecological response to modern environmental challenges, offering a new perspective on life, death and the natural world.

These projects highlight the evolution of interaction design as it moves beyond the digital and mechanical to embrace the living, offering new possibilities for sustainability and deepening our relationship with the natural world.

Lighting Design

Lighting design plays a central role in shaping human experience, influencing well-being and contributing to the health of the environment. It defines atmospheres, enhances aesthetics and optimizes functionality in various fields, including architecture, urban planning and interior design. However, the quest for sustainability and energy efficiency in lighting presents unique challenges, particularly in addressing the environmental impact of modern lighting technologies.

The advent of LED technology has brought significant energy savings and reduced carbon footprints. However, its widespread adoption has also raised environmental concerns, particularly in relation to manufacturing processes, disposal challenges and the potential inclusion of hazardous materials (Martins et al., 2020). This dichotomy has led to a renewed interest in nature-inspired lighting solutions in line with sustainable practices.

Historically, humankind has relied on natural sources of light such as fire and bioluminescence. Fire, as the primary source of illumination for millennia, symbolized human ingenuity in harnessing nature for practical use. Meanwhile, bioluminescence – observed by Darwin and exploited by early societies through substances derived from organisms such as fireflies – represented an early example of a symbiotic relationship with the environment (Campbell, 2012; Srivastava & Katiyar, 2021). Substances such as luciferin and the enzyme luciferase, derived from bioluminescent organisms, facilitated the development of innovative lighting practices long before artificial light revolutionized human interaction with lighting (Oba et al., 2020; Jabr, 2016; Johns, 2021).

In today's search for environmentally friendly solutions, scientists are revisiting ancient and natural sources of light, such as bioluminescence, as an alternative to conventional artificial light (Barati et al., 2021). This phenomenon, characterized by the emission of light by living organisms, occurs through the oxygenation of a substrate called luciferin, catalyzed by the enzyme luciferase. Unlike fluorescence and phosphorescence, bioluminescence generates light without relying on the absorption of external electromagnetic radiation (Kahlke & Umbers, 2016; Wilson, 2013). Its diverse ecological roles, from mate attraction to predator deterrence, underscore its evolutionary importance and potential for human applications.

Bioluminescence has far-reaching implications for sustainable lighting. Its potential ranges from biomedical research, where it helps to understand cellular processes (Sharifian et al., 2018; Paley & Prescher, 2014; Mali, 2024), to pioneering applications in environmentally friendly and innovative lighting systems. These possibilities exemplify the principles of DILS, where living organisms are integrated into functional and regenerative design solutions. DILS approaches could transform lighting design by embedding bioluminescent systems into urban and architectural environments, offering energy-efficient alternatives that redefine human interaction with light.

The integration of bioluminescence into DILS represents a convergence of biology, design and sustainability. By drawing inspiration from nature, lighting design can evolve towards solutions that are not only environmentally responsible but also enhance human engagement with the living systems that shape our world. This approach fosters a deeper connection between technological innovation and ecological harmony, paving the way for a brighter, more sustainable future in lighting design.

Glowee, a pioneering start-up company, has introduced an innovative approach to urban lighting with the launch of its bioluminescent streetlights in Rambouillet, France. This initiative is driven by the dual goals of reducing energy consumption and reducing light pollution. As part of its efforts to engage the public in this cutting-edge technology, Glowee initiated the Glowpolis[41] project, which involved the creation of furniture and street lighting illuminated by bioluminescent bacteria. Building on this foundation, the company has expanded its offerings to include GlowEvents,[42] immersive installations that combine bioluminescent art with event décor; and Biolumina,[43] an educational board game that teaches players about the science behind natural and synthetic bioluminescence.

Glowee's work illustrates the potential of bioluminescence as a sustainable and interactive lighting solution. In the context of DILS, these applications not only highlight the integration of biological systems into urban design but also underline the importance of fostering symbiotic relationships between living organisms and human-built environments. The incorporation of bioluminescent bacteria into public infrastructure demonstrates how DILS can address ecological concerns while enriching human interaction with biological systems.

Future developments in bioluminescent technology will go beyond street lighting and pave the way for living, interactive lamps. These innovative solutions could use bioluminescent bacteria or algae to provide sustainable, dynamic alternatives to conventional lighting methods. The potential of such systems lies in their ability to create responsive and energy-efficient lighting that adapts to environmental and human interactions.

A remarkable example of this visionary application is the Growing Light[44] project by Tim van Dortmont, developed as part of his final project at TU Delft under the supervision of Elvin Karana and Bahar Barati (see Figures 3.18, 3.19 and 3.20). Exploring the dynamic interplay between living organisms and user interaction, this project presents an installation of cubes containing orbital vibrators and cultures of bioluminescent algae. Equipped with motion sensors, the cubes respond to the presence of visitors by gently shaking, which activates the algae to emit light. The intensity and patterns of agitation evolve as visitors approach, creating a range of light behaviors. After a series of interactions, the cubes pause to allow the algae to recover, emphasizing the biological nature and limitations of the material (van Dortmont, 2019; Barati et al., 2021).

In addition to demonstrating the diverse potential of bioluminescent organisms in design, the Growing Light installation embodies the principles of DILS by fostering a deeper understanding of and connection to living systems. By showcasing the responsive capabilities of algae, this project bridges the gap between biological processes and human-centered design, highlighting the transformative potential of integrating living organisms into functional and aesthetic applications.

Designed by van Dongen as a final project at the Design Academy Eindhoven, the Ambio Lamp[45] exemplifies the innovative use of bioluminescence in lighting applications (Figure 3.21). This cutting-edge lamp consists of a glass tube filled with a saline solution, which supports bioluminescent bacteria extracted from the skin of an octopus. Once isolated and cultured, the bacteria are introduced into a synthetic seawater medium enriched with nutrients essential for their survival and luminescent activity.

Figure 3.18 Growing Light (2020) prototype overview by Tim van Dortmont. The setup features two metal bases with integrated rotary shakers, each supporting a sealed glass vessel containing a living culture of *Pyrocystis fusiformis*.

Credits: Tim van Dortmont.

Figure 3.19 Growing Light (2020). Pulse agitation of a *Pyrocystis fusiformis* culture within the installation, illustrating the dynamic interaction with the bioluminescent medium.
Credits: Tim van Dortmont.

Figure 3.20 Growing Light (2020). Top view of a *Pyrocystis fusiformis* culture after the application of rotational force, showcasing the resulting bioluminescent response.
Credits: Tim van Dortmont.

The lamp's design incorporates a balanced structure with a glass tube on one side and a counterweight on the other, allowing for a gentle rocking motion. This rocking motion oxygenates the solution and activates the bioluminescent properties of the bacteria. With a minimal push, the lamp emits a sustained glow for approximately 20 minutes, demonstrating the potential of bioluminescence as an

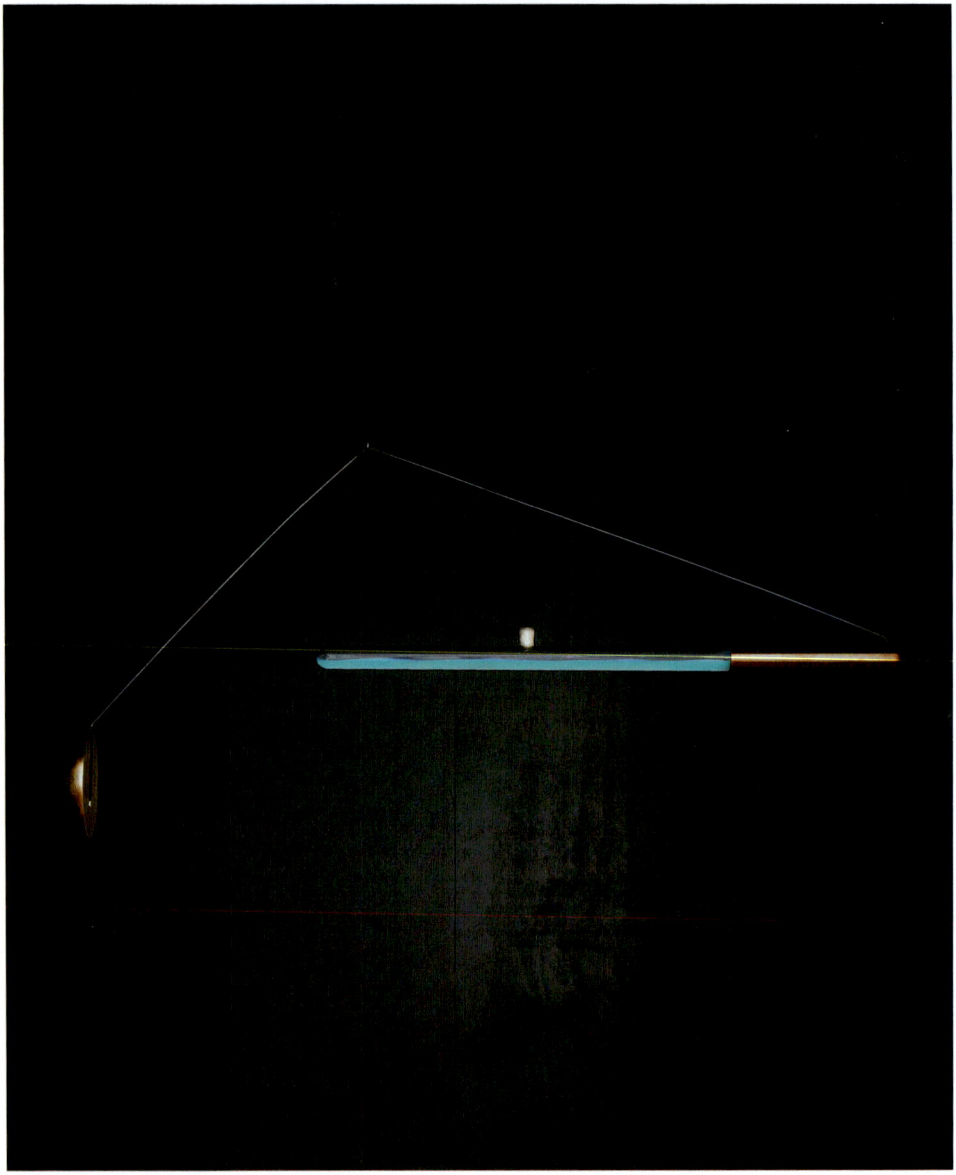

Figure 3.21 Ambio Lamp (2014) by Teresa van Dongen. An innovative lighting design powered by the metabolic processes of marine bioluminescent bacteria, exploring the potential of living organisms as sustainable light sources. Credits: Roel van Tour. Courtesy: Teresa van Dongen.

environmentally friendly alternative to traditional lighting sources (Hobson, 2015).

The Ambio Lamp is closely aligned with the principles of DILS and provides a tangible example of the symbiosis between design, living organisms and sustainable technology. By incorporating living organisms directly into the design process, the lamp not only redefines the aesthetics of lighting, but also promotes a shift towards regenerative and environmentally conscious design practices. The interplay of movement and light in this system emphasizes the dynamic relationship between the user and the living organism within the design, highlighting the transformative potential of DILS in creating sustainable and interactive solutions for everyday life.

The search for ways to illuminate spaces through bioluminescence has spurred significant advances in synthetic biology, leading to the development of ornamental plants capable of emitting natural light. A prominent initiative in this area is the collaboration between Light Bio Inc.,[46] a pioneer in bioluminescent plants for the home and landscape, and Ginkgo Bioworks,[47] a leading US-based synthetic biology laboratory. The partnership aims to increase the luminosity and efficiency of the enzymes in Light Bio's glowing plants by leveraging Ginkgo's advanced plant cell engineering capabilities. Light Bio has successfully adapted a bioluminescence system derived from luminous fungi to develop plants that glow with remarkable brightness and efficiency. These advances exemplify the potential of synthetic biology to integrate living systems into everyday applications, in line with the principles of DILS, by fostering sustainable, biointegrated innovation.

The search for innovative lighting solutions extends beyond bioluminescent plants to the creative intersection of microbiology and design. At the Young Swedish Design Exhibition in Stockholm, designer Jan Klingler presented the Bacteria Lamp,[48] a unique LED lamp that incorporates bacterial cultures into its aesthetic and functional design. This groundbreaking work bridges industrial design and microbiology, with two lamp styles: one resembling a laboratory flask and another as a circular wall-mounted piece, both containing living bacterial cultures encased in resin discs. Inspired by laboratory techniques, Klingler cultivates a variety of bacteria, yeasts and fungi on agar substrates within the resin discs. The vibrant colors of the lamps are derived either from the natural pigmentation of the microbes or from the growth medium itself.

Klingler's meticulous manufacturing process involves pouring resin into molds, embedding a silicone ring and incorporating the cultured microbial cultures. Once the desired visual effect is achieved, he stops the bacterial growth by sealing the disc and cutting off the oxygen supply, effectively creating what he describes as "modern fossils".[49] The addition of LED lighting to the resin discs accentuates the intricate microbial patterns, creating a fascinating glow that highlights the intersection of natural processes and contemporary design.

Fashion Design

The fashion industry has long been a major contributor to critical environmental problems, promoting a throwaway culture and exacerbating ecological challenges. Fast fashion, characterized by rapid production and disposal cycles, has resulted in significant environmental and social costs (Cobbing & Vicaire, 2016; Niinimäki et al., 2020; Jacometti, 2019). These include water pollution, high greenhouse gas emissions from textile production, extensive waste generation and the widespread use of synthetic fibers, which contribute to microplastic pollution and hinder material recovery at the end of their lifecycle.

In response to these pressing challenges, the fashion sector is actively seeking sustainable solutions to reduce the environmental impact of textile production and waste management throughout the supply chain. One of the most promising approaches is the integration of biobased, biofabricated and living materials, which offer viable alternatives to conventional textiles (Seidu et al., 2023; Lee et al., 2020; Ahmad, 2020; Collet, 2021; Rognoli et al., 2022; Quijano et al., 2021). These innovations are paving the way for the development of textiles derived from novel sources such as mycelium and bacterial microcellulose (Amobonye et al., 2023). Such materials contribute to the creation of active and adaptive fabrics that can perform various functions relevant to human needs and environmental challenges (Provin et al., 2021; Rathinamoorthy et al., 2023).

Beyond sustainability, these biofabricated and living textiles exemplify the principles of DILS, which merge biological systems with advanced design methodologies. DILS in fashion not only enhance the functionality of fabrics but also open up new ways of creating vibrant colors, intricate textures and enhanced material properties through biological processes. This approach fits into a broader vision of embedding dynamic and interactive elements into the very fabric of design, transforming garments into systems that adapt, respond and integrate seamlessly with their environment.

While many of these innovations are still at an experimental or speculative stage, ongoing research continues to lay the foundations for practical applications in sustainable fashion. The integration of DILS represents a paradigm shift, offering not only a reduction in the environmental footprint of the fashion industry, but also a redefinition of consumption practices towards a responsible and sustainable engagement with materials. By embracing the potential of living systems and eco-friendly textiles, the fashion industry is taking significant steps towards a future where design is in harmony with environmental stewardship and regenerative principles.

The integration of bioluminescence into fashion represents a pioneering approach to enhancing aesthetic appeal while exploring sustainable design practices. One notable example is the Living Light Dress,[50] a pioneering project developed in 2017 through a collaboration between designer Victoria Geaney and scientists Anton Kan and Bernardo Pollak from the University of Cambridge.[51] This innovative endeavor merges biology and design by infusing garments with the natural glow of bioluminescent bacteria, pushing the boundaries of traditional fashion.

The research team undertook a meticulous process to identify suitable materials capable of harboring bioluminescent bacteria. They experimented with a range of fabrics, including nylon, cotton, wool and felt, and eventually developed a seawater-based agar jelly enriched with amino acids, carbohydrates and yeast extract to feed the bacteria. This nutrient-rich medium supported both the growth and luminescence of the microorganisms.

To impregnate the fabric with this medium, the team immersed the dress in the prepared agar solution and allowed it to dry. Once the nutrient-filled fabric was ready, they introduced *Photobacterium kishitanii*, a bioluminescent bacterium originally isolated from a deep-sea fish by Japanese scientist Henryk Urbanczyk (Ast et al., 2007; Dunlap & Urbanczyk, 2013). This species is known for its ability to emit natural light under certain conditions, making it an ideal candidate for the project.

The Living Light Dress is an example of the integration of DILS into fashion (see Figures 3.22 and 3.23). By embedding living organisms into textiles, this project not only achieves a dynamic and interactive aesthetic, but also highlights the potential for environmentally friendly and innovative clothing. It reimagines garments as active, living systems that respond to their environment and user interactions, in line with wider trends towards sustainability and regenerative design.

This project is more than an experiment in bioluminescent fashion; it represents a significant leap in the interdisciplinary exploration

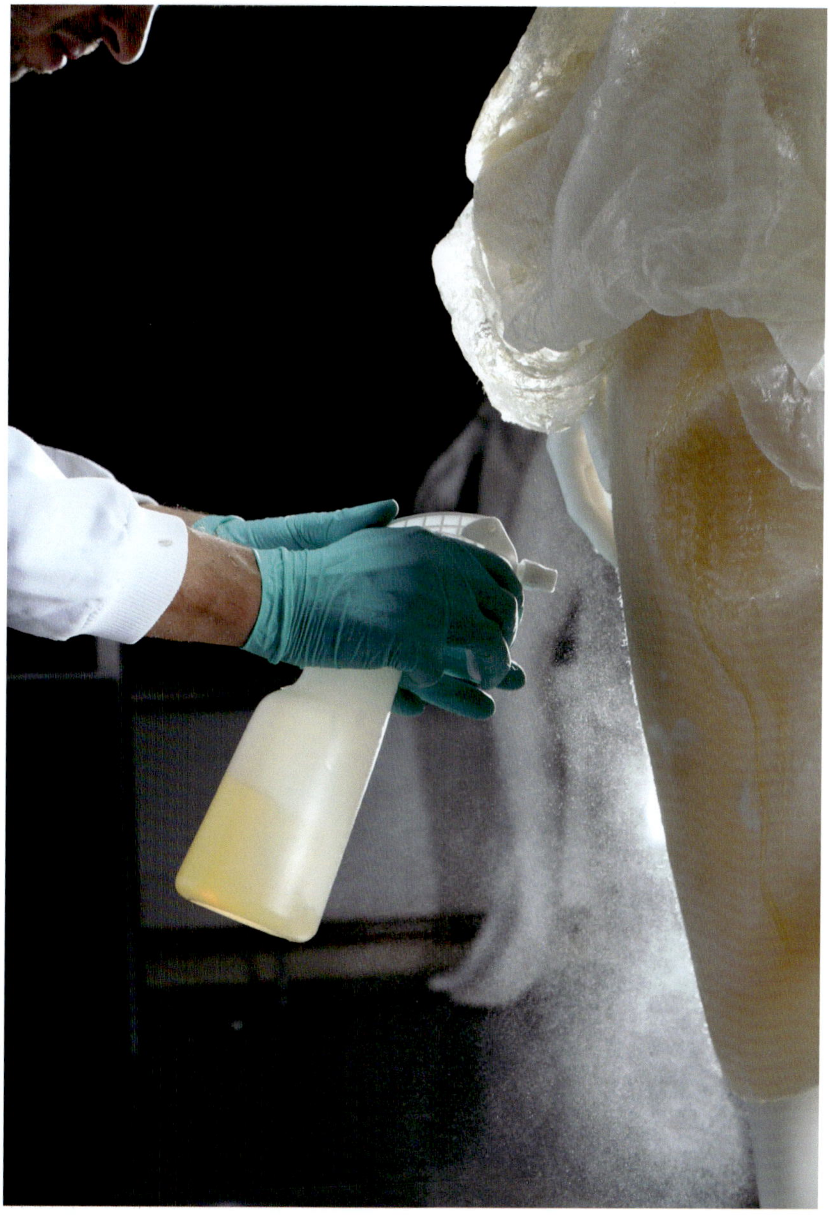

Figure 3.22 Living Light Dress (2017), preparation process. Laboratory view showing the application of *Photobacterium kishitanni* cultures onto the biopolymer textile surface during the development of the bioluminescent dress.
Credits: Photography by Chris Hoare. Design by Dr. Victoria Geaney, Dr. Bernardo Pollak, and Dr. Anton Kan.

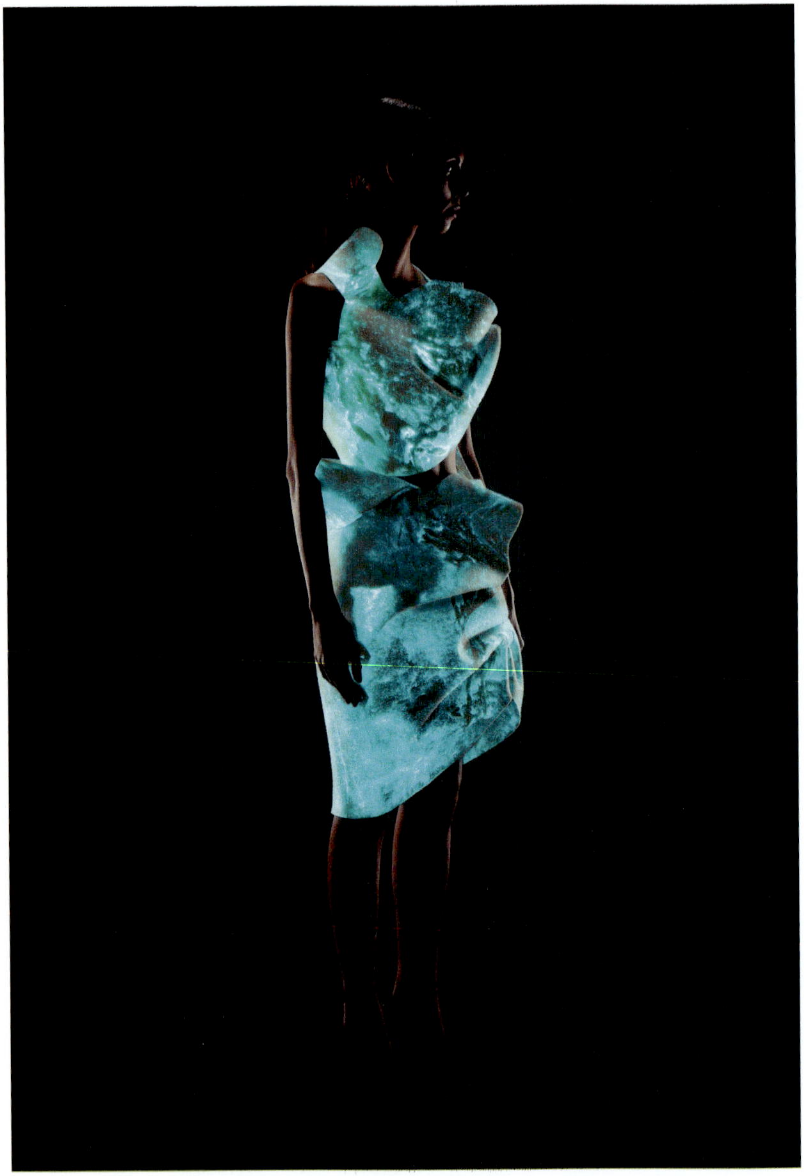

Figure 3.23 Living Light Dress (2017), final prototype. A bioluminescent garment incorporating *Photobacterium kishitanni*, demonstrating the interaction between living organisms and wearable technology.

Credits: Photography by Chris Hoare. Model: Manuella Gomide (D1 Models). Hair and make-up by Victoria Winfield. Published in *Wired* magazine, London, February 2017 issue.

of biology and design. It paves the way for future innovations that combine living systems with textiles and offers a vision of the possibilities of sustainable, interactive and aesthetically transformative clothing for the fashion industry.

Lucid Life,[52] an innovative project by designer Christopher Bellamy, exemplifies the integration of bioluminescence into fashion and cultural artifacts. Inspired by the natural symbiosis between coral and microorganisms, Bellamy developed a groundbreaking living material in 2024 that incorporates bioluminescent microalgae. This material remains viable for at least 5 months, responds to touch with a soft glow and requires only sunlight for maintenance. To explore the potential of this material in fashion and cultural design, Bellamy worked with Polynesian artisans, combining advanced scientific techniques with traditional craftsmanship. This collaboration resulted in a collection of artifacts that seamlessly blend living bioluminescent materials with indigenous practices and locally sourced resources. Notable creations include a bioluminescent swimsuit that combines functionality with a luminous aesthetic, demonstrating the adaptability of the living material to wearable applications. The garment embodies the dynamic interplay of human interaction, light and living systems, and demonstrates the potential for creating interactive and responsive designs. Another significant artifact was a luminous necklace that integrates the living material into traditional Polynesian jewelery-making practices. This piece captures the intersection of cultural heritage and ecological innovation, symbolizing the ephemeral beauty and vibrancy of bioluminescence.

The artifacts produced through Lucid Life reflect the inherent fragility and impermanence of DILS, resonating with the ephemeral nature of Polynesian traditional knowledge that exists only in living memory. While the ephemeral nature of the material introduces vulnerability, it also fosters adaptability, responsiveness and interaction, creating an intimate connection between the wearer and their environment.

By integrating DILS into cultural and fashion contexts, Lucid Life highlights the transformative potential of living materials in promoting sustainable, dynamic and culturally resonant designs. This project underscores the importance of interdisciplinary collaboration in advancing the applications of living systems and offers a glimpse into a future where fashion and cultural artifacts embody ecological innovation and symbiotic relationships with nature.

The integration of living organisms into clothing represents a transformative approach to interactive design, particularly in the field of human health. By augmenting the wearer's natural microbiome with beneficial bacteria, these systems open up new possibilities for wearable technology that is both functional and sustainable. A notable example of this innovation is Skin II, a project developed by Central Saint Martins fashion designer Rosie Broadhead in collaboration with microbiologist Christopher Callewaert.[53] This project uses probiotics activated by the wearer's perspiration and strategically embedded in moisture-prone areas of the garment to create a symbiotic relationship between the skin and the fabric.

This innovative application of DILS demonstrates how the body's natural processes can be used to enhance garment functionality while minimizing environmental impact. By eliminating the need for chemical treatments and reducing the frequency of laundering, Skin II is in line with the principles of sustainable fashion (Broadhead et al., 2022; Broadhead et al., 2021). The probiotics incorporated into the fabric promote a dynamic interaction with the skin, providing benefits such as improved skin health and odor reduction. Importantly, this interaction is achieved without the need for electrical components or external power sources, relying solely on the natural behavior of the microorganisms.

The incorporation of living organisms into a leotard-like fabric represents a significant breakthrough in wearable probiotic technology and illustrates the potential of DILS to redefine the relationship between clothing, the human body and the environment.

By fostering a deeper connection between the wearer and their microbiome, this project offers a new paradigm in clothing design – one that prioritizes ecological innovation and human well-being, while challenging traditional notions of textile functionality. This pioneering approach not only exemplifies the potential of DILS in fashion but also highlights its role in advancing sustainable and health-conscious design practices.

The integration of living organisms into textiles is reshaping the fashion industry, bringing together biology, design and technology to create dynamic, sustainable and interactive solutions. In this context, the BioLogic[54] project, developed by the Tangible Media Group at MIT Media Lab (Figure 3.24), exemplifies the transformative potential of DILS in fashion. Harnessing the hygroscopic properties of *Bacillus subtilis natto*, this project developed a responsive

Figure 3.24 BioLogic Project (2015), close-up view. Detail of the biohybrid textile surface that dynamically reacts to the wearer's perspiration.
Credits: Rob Chron.

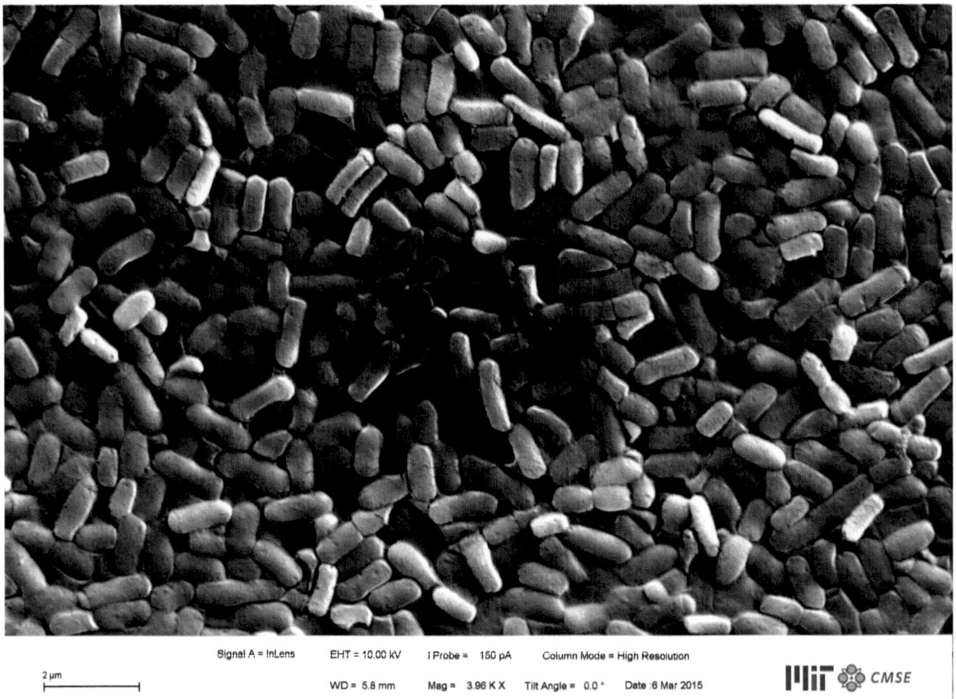

Figure 3.25 BioLogic Project (2015), microscopic view. Scanning electron microscope (SEM) image showing the structure of the biohybrid film incorporating *Bacillus subtilis natto* bacteria, developed at the MIT Center for Materials Science and Engineering (CMSE).
Credits: Rob Chron.

"bio-skin" that adapts to the wearer's body heat and perspiration (Figure 3.25). High precision bioprinting was used to embed adaptive textile flaps into the garment, which dynamically open and close to promote sweat evaporation and cooling (Yao et al., 2015). This interaction between fabric and physiology introduces a new dimension to clothing, where garments not only serve aesthetic and functional purposes, but also respond to the body's needs to enhance comfort and thermal regulation.

Another significant milestone is Roya Aghighi's Biogarmentry[55] project, which integrates algae into wearable fabrics to create biodegradable textiles capable of cleaning the air through photosynthesis. Working with scientists at the University of British Columbia, Aghighi explored embedding photosynthetic cells into natural fibers to create garments that contribute to the health of the environment while requiring active care from the wearer. This interactive relationship between the wearer and the algae highlights the role of DILS in fostering sustainability and promoting mindful consumption practices (Aghighi, 2019; Ertürkan et al., 2022). The biogarment initiative not only redefines textile functionality but also highlights the potential of DILS to merge human care with environmental stewardship.

The fashion industry is increasingly exploring biodesign as a transformative approach to sustainability and innovation, with major players integrating living systems into their offerings. In a pioneering venture in 2017, Puma, in collaboration with Biorealize[56] and the MIT Design Lab, embarked on a project to explore the future of athletic gear by incorporating living organisms to enhance performance.[57]

The collaboration resulted in the unveiling of four diegetic prototypes[58] at Milan Design Week 2018, demonstrating the vast potential of biological design and manufacturing in sportswear. These innovations included a "breathing" shoe that can customize its ventilation by maintaining its own air channels, a deep learning insole designed to enhance athletic performance through real-time biofeedback from organisms that detect chemical signs of fatigue or health, a microbially responsive T-shirt that adapts its appearance based on environmental conditions and provides feedback on air quality, and a revolutionary packaging material. This biologically engineered material can change its shape and structure, offering a new paradigm for living, biodegradable and adaptable packaging.

These breakthroughs highlight the intersection of biology, design and technology, and offer a glimpse of a future where clothing and accessories not only meet sustainability criteria, but also engage in a dynamic interaction with their wearers and the environment.

The Amoeba trainer is a pioneering project in sports footwear, the result of a collaboration between designer Shamees Aden and scientist Dr Martin Hanczyc. Designed for the year 2050, the project harnesses the innovative potential of protocells – which blur the lines between the living and non-living – to revolutionize materials development.[59]

Protocells, or cell-like entities (CLEs), are formed from inert liquids and represent a fusion of living and non-living elements. This collaboration uses 3D-printed biotechnology to create a "second skin" around the wearer's foot, offering seamless, responsive adaptation to the individual's movements. The protocells dynamically adapt to provide targeted support during different activities, demonstrating a unique blend of adaptability and responsiveness.

While protocells exhibit behavior like living organisms – such as environmental adaptation, movement and self-organization – they do not meet the criteria for living organisms as they cannot reproduce or evolve. Emerging from bottom-up, self-organizing processes, protocells and CLEs mark a departure from the top-down manipulation of life of traditional synthetic biology.

Current research into protocells focuses on creating advanced biosensors that can detect changes in the environment. However, the broader implications of protocell research extend far beyond material production, promising to significantly impact future design paradigms and offering a glimpse into a world where the boundaries between the living and the non-living are increasingly blurred (Marenko, 2018).

Charlotte Furet's project Chemical Synchronisation[60] presents an innovative approach to enhancing human interaction by borrowing from the natural communication methods of animals, plants and insects, which rely on subtle chemical signals. This project introduces a wearable device, similar to a necklace, that enables the wearer to sense and harmonize with the emotional states of those around them. Developed in collaboration with Takeuchi Lab, the device is equipped with a sophisticated sensor with chemoreceptors similar to those found in insects, enabling it to identify specific emotions such as happiness and enthusiasm by analyzing chemical markers in human sweat. Having detected these emotions, the device can amplify and share them with others through a microneedle patch that delivers hormones such as oxytocin and endorphins to promote a shared emotional experience.

Looking to the future, the team behind Chemical Synchronisation envisions expanding the device's capabilities to detect a wider range of emotions and facilitate chemical communication over distance, potentially revolutionizing the way people connect with each other.

Visual Communication

The field of visual communication has witnessed significant change and diversification, driven by the emergence of new technologies and the proliferation of media platforms. This evolution has been characterized by the coexistence of digital and traditional analog media, offering different ways of representation, expression and interaction. The interplay between these media has expanded the boundaries of visual communication, providing a richer palette for creative and functional applications.

Historically, visual communication has relied heavily on analog forms such as paintings, drawings, prints and photographs, which dominated the pre-digital era. With the technological advances of the late twentieth century, digital visual communication emerged as a pivotal force, encompassing animated graphics, video, digital art and immersive virtual reality applications. Whether analog or digital, visual communication plays an essential role in facilitating human interaction and enhancing the ability to effectively convey messages, emotions and information (Adiloglu, 2011). In an age of information overload and diminishing attention spans, compelling visual communication has become essential to create powerful and memorable impressions.

As visual communication becomes seamlessly integrated into everyday life, its environmental impact warrants critical examination (Aydemir & Özsoy, 2020). Both analog and digital forms have distinct environmental footprints that are shaped by their production, distribution and consumption processes.

The environmental footprint of analog visual communication is closely linked to the extraction of raw materials, manufacturing processes and transport of physical media such as paper, billboards, inks and paints. These materials contribute to resource depletion, carbon emissions and pollution. For example, the production of paper and inks involves deforestation, the use of chemicals and energy-intensive processes, all of which contribute significantly to environmental degradation.

Digital visual communication, often perceived as a sustainable alternative, presents its own environmental challenges (Malmodin & Lundén, 2018; Zhou et al., 2019). The exponential growth in digital content consumption has led to an increased demand for data storage and processing capacity (Batmunkh, 2022). Data centers, which power the digital ecosystem, consume large amounts of electricity, much of which comes from non-renewable sources. This energy consumption contributes significantly to the carbon footprint of digital visual communication (Belkhir & Elmeligi, 2018). In addition, the rapid obsolescence of the electronic devices used to create and consume digital content exacerbates the problem of e-waste. These devices often contain hazardous materials, and the challenges of recycling them pose significant environmental risks. The inability to effectively manage e-waste contributes to the growing environmental impact of the digital visual communications industry.

The integration of DILS into visual communication offers a transformative approach to addressing these environmental concerns. By incorporating living organisms, biobased materials and sustainable processes into the creation of visual media, DILS can redefine the impact of visual communication.

For example, innovations such as algae-based bioluminescent displays or bacteria-driven pigments could provide sustainable alternatives to conventional digital screens and analog printing methods. These solutions are in line with the principles of promoting energy efficiency, reducing resource consumption and minimizing waste. DILS also offer the potential for interactive and dynamic visual media that actively engage with their environment, creating a deeper connection between audience, medium and nature.

As visual communication continues to evolve, embracing the principles of sustainability and integrating DILS into its practices could lead to a paradigm shift. This approach not only addresses the environmental challenges posed by traditional and digital media, but also provides new opportunities for creative expression, fostering a harmonious relationship between design and the more-than-human world.

Bioluminescence, with its intriguing natural glow, is emerging as a groundbreaking innovation in the field of visual communication, offering new ways of presenting and disseminating information. A notable example of this development is French designer Charlotte Furet's Living Pixels[61] project. By integrating bioluminescent bacteria into the realm of data visualization, this project introduces a paradigm shift in how information can be displayed and experienced.

The Living Pixels project uses bioluminescent microorganisms to create an interactive, living display capable of visualizing data in a distinctively organic way. By moving away from the predictability of conventional digital screens, the project explores the integration of biological rhythms and environmental responsiveness into the communication process. The bioluminescent bacteria used in the installation emit light based on their metabolic processes, introducing an element of unpredictability and wonder into the exchange of information.

This project critiques the rigidity of standard modes of information dissemination, where static displays dominate, by proposing a dynamic and interactive alternative. The living display responds to human interaction while adapting to environmental changes such as temperature and humidity, reflecting the delicate interplay between human activity and natural systems. This approach not only enhances engagement, but also evokes the aesthetics of natural ecosystems, creating a multisensory experience that connects the viewer to the living materiality of the medium.

As a DILS, Living Pixels is an example of how living organisms can transform visual communication. DILS enable the seamless fusion of biological and computational systems, offering the potential for sustainable, adaptive and emotionally engaging interactions.

Through projects like Living Pixels, the integration of bioluminescence into visual communication not only expands the possibilities for data representation, but also fosters a deeper connection between audiences and the living world. This innovative use of natural processes highlights the potential of DILS to redefine visual media, making it more dynamic, engaging and environmentally conscious.

Building on the innovative foundation of Living Pixels, other groundbreaking projects continue to demonstrate the immense potential of bioluminescence for visual communication. One important example is the development of Bioluminescent Bacterial Billboards[62] by Dr Jeff Hasty and his research team at the University of California, San Diego. This pioneering work, carried out within the Division of Biological Sciences and the BioCircuits Institute, demonstrates how living systems can be engineered and designed to produce synchronized and visually striking displays of light. These "living neon signs" consist of engineered *E. coli* bacterial cells programmed to emit synchronized, intermittent flashes of light. The underlying technology uses a sophisticated mechanism to regulate the biological rhythms of thousands of bacterial cells, enabling coordinated light production across entire colonies (Prindle et al., 2012; Danino et al., 2010). The result is a dynamic and fascinating display that demonstrates the aesthetic potential of synthetic biology and also highlights its ability to precisely control living cells as if they were programmable machines.

The practical applications of this technology go beyond visual appeal. By exploiting the biological rhythms that govern these bioluminescent displays, the researchers have developed functional tools such as bacterial sensors. For example, a simple biosensor developed by the team uses the blinking patterns of bacteria to detect and quantify traces of arsenic. The rhythmic flashing serves as an indicator of the presence and concentration of the toxin, illustrating the bacteria's inherent responsiveness to environmental stimuli.

The implications of this research are profound, suggesting a future where living biosensors could be used for a wide range of environmental monitoring applications. Thanks to bacteria's innate sensitivity to specific substances, these biosensors could be adapted to identify various pollutants and pathogens in water, soil or air. Unlike conventional chemical sensors, the living nature of these biosensors allows them to respond dynamically to changes in toxin concentrations over time, offering a more adaptable and sustainable solution.

Artist and scientist Hunter Cole[63] bridges the disciplines of art and science through a diverse body of work that includes painting, photography, digital art and an extraordinary focus on bioluminescent living art. By harnessing the natural glow of bioluminescent bacteria, Cole creates living drawings that embody the eternal cycle of life and death. These works offer a contemplative reflection on human mortality and the passage of time[64] (Cole, 2018). Her work is widely recognised for merging scientific enquiry with aesthetic exploration, as discussed by Dybas (2019) in Oceanography, who highlights Cole's unique integration of marine biology and artistic practice.

Cole's innovative artistic process involves cultivating bioluminescent bacteria in petri dishes to form intricate line drawings. These bacteria, an essential narrative element, emit a bright, natural glow as they metabolize nutrients. The works are photographed at their peak luminescence to capture their vibrant, ephemeral beauty. Over the course of approximately 2 weeks, the bacteria's glow gradually fades as they deplete their resources and expire naturally. This transformation underscores the transient nature of life, making the artwork a dynamic, vibrant representation of impermanence.

Expanding the sensory experience, Cole has documented these luminous works of art in a film that captures their evolution from vibrant life to quiet silence. The auditory dimension of her work is equally innovative; the film's soundtrack is composed of the protein sequences of the bacteria themselves, merging the scientific and artistic realms into a unique multisensory experience.

One of Cole's most celebrated projects, Living Light, further explores the interplay between bioluminescence and human experience. In particular, her Bioluminescent Weddings series shows couples in wedding dresses illuminated by the soft, ethereal glow of bioluminescent bacteria (Figure 3.26). The series is inspired by nature, where bioluminescence often functions as a

Figure 3.26 Living Light Project, "Bioluminescent Weddings" series (2017). Angel Bride – photograph illuminated solely by the light of bioluminescent bacteria, part of Hunter Cole's ongoing exploration of living light as both artistic and biological medium.
Credits: Hunter Cole.

means of attraction, symbolizing unity and connection. Combining human rituals with the enchanting qualities of living light, the project highlights the aesthetic and symbolic potential of bioluminescence (Figure 3.27).

Cole's work exemplifies the profound possibilities of DILS in visual communication and art. Her work is a powerful example of how art and science can converge to create evocative, vibrant experiences that challenge conventional perceptions of both disciplines.

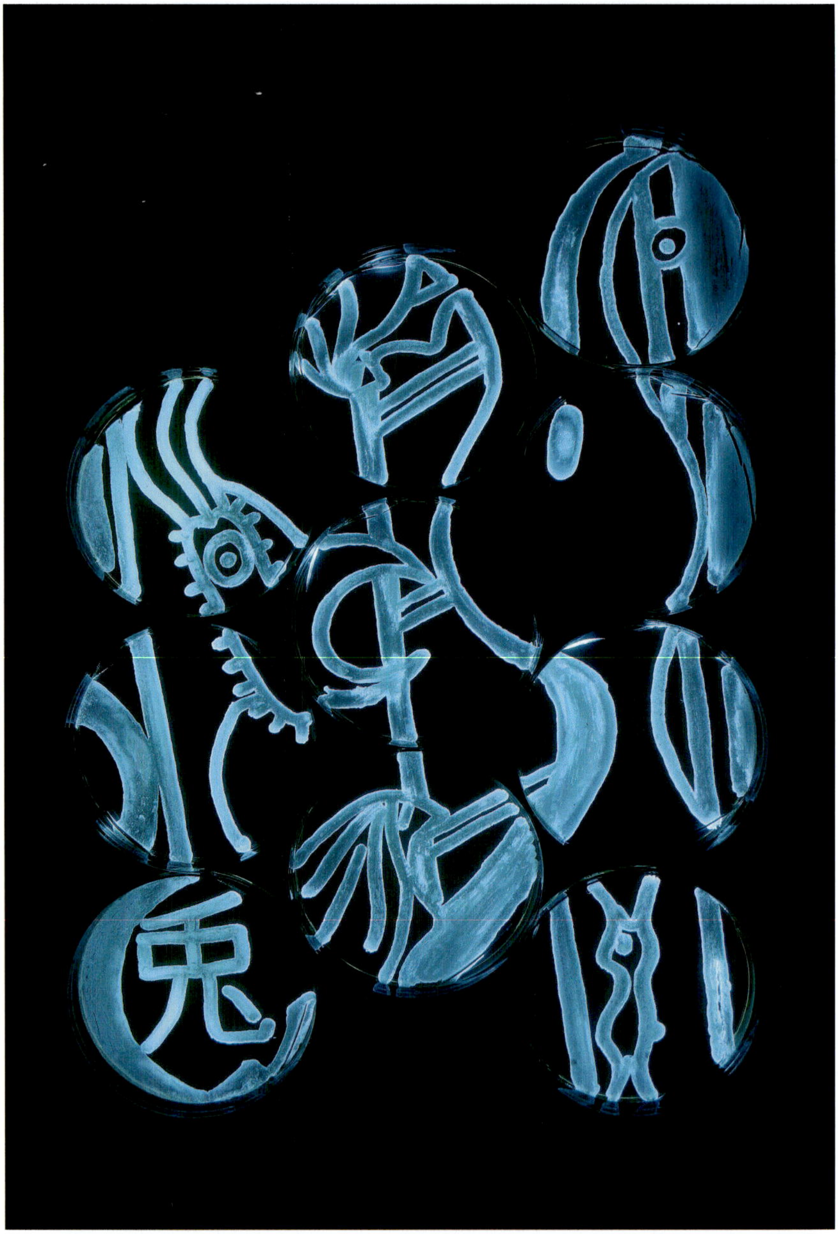

Figure 3.27 Living Light Project (2017), "Rabbit: Stage 1" – photograph illuminated solely by the light of bioluminescent bacteria, part of Hunter Cole's ongoing exploration of living light as both artistic and biological medium.

Credits: Hunter Cole.

In the world of advertising, the Contagion poster developed by Warner Bros. Pictures Canada[65] is a compelling example of how the integration of science and design can lead to marketing innovation. This groundbreaking campaign, created to promote Steven Soderbergh's film *Contagion*, fused microbiology and visual communication to generate strong public engagement and response. Working with mycologist Patrick Hickey,[66] the project used live microorganisms to create an unconventional microbial billboard that was prominently displayed in the window of an empty storefront at 409 Queen Street West in Toronto. Central to this creative project was the use of live bacteria, including *Penicillium*, mold and colorful microbes, encapsulated in two large petri dishes. These petri dishes acted as the substrate for a dynamic, living display that vividly symbolized the theme of viral spread depicted in the film's storyline. Over time, the bacteria and mold visibly grew, transforming the billboard and providing passers-by with a real-time visualization of microbial proliferation. This interactive, evolving medium effectively immersed the audience in the film's narrative, making the campaign both thought-provoking and scientifically engaging (Travis, 2011).

This innovative advertising strategy exemplifies the potential of DILS to revolutionize public engagement in marketing campaigns. By utilizing the dynamic and evolving qualities of living organisms, the Contagion poster transformed a traditional static advertisement into an interactive, living work of art. This approach not only highlighted the theme of contagion in a tangible way, but also bridged the gap between science, design and popular culture.

The Electric Factory, a company renowned for its creative approach to brand challenges,[67] recently launched a groundbreaking campaign for soap brand Lifebuoy called Bacteria ADS. The campaign featured a visually engaging installation consisting of a large petri dish filled with live colonies of bacteria juxtaposed with commonly used items such as smartphones, money and game controllers. The dynamic display was designed to vividly demonstrate the ubiquitous presence of bacteria and fungi on these everyday objects, thereby highlighting the importance of hand hygiene.[68] Through this innovative approach, The Electric Factory effectively communicated the omnipresence of microorganisms in everyday life and highlighted the critical role of cleanliness in preventing the spread of germs. This campaign is an example of how DILS can be used in marketing to create powerful, thought-provoking messages that combine science with public awareness.

Beyond marketing, the integration of algae and bacteria into visual media is redefining artistic practices, particularly in the field of photography. By incorporating living microorganisms into the photographic process, artists are drawing attention to the transient nature of human existence, juxtaposed with the enduring quality of images that capture life's moments. This innovative intersection of art and biology aligns with scientific advances in understanding the photobiological capabilities of algae and bacteria.

In particular, microalgae and cyanobacteria exhibit highly complex light-sensing systems, including photoreceptors and pathways that are influenced by light stimuli (Kianianmomeni & Hallmann, 2014). These systems can regulate essential biological functions such as photosynthesis and also provide a basis for exploring the interaction between light and living systems. Artists are increasingly using these properties to create dynamic, living photographs and installations that evolve over time, reflecting the interplay between impermanence and permanence inherent in both biological systems and the photographic medium.

This fusion of art and science highlights the versatility of DILS in expanding creative boundaries and providing new lenses through which to explore fundamental concepts of life, decay and regeneration.

A pioneering example of this new technique is Lia Giraud's[69] Cultures project, which introduces the innovative technique of algaegraphy,[70] – a method that uses motile, light-sensitive microorganisms, specifically microalgae, to create living images. This groundbreaking work exemplifies the potential of DILS to merge biological processes with artistic expression.

Giraud's algaegraphy involves a meticulously crafted process that unfolds in three sealed units, each containing a petri dish with a culture of microalgae. A negative image is projected onto the dishes, guiding the light-sensitive microalgae to align themselves within the illuminated regions optimal for their growth. Over a 4-day incubation period, the microorganisms self-organize and adjust their density to form an intricate living image. The process concludes with the solidification of the medium, which encapsulates the algae in a nutrient-rich substrate that supports their growth. Once complete, the living works of art are removed from the incubator and displayed on illuminated tables, allowing the public to witness the dynamic, evolving nature of the microalgae as they continue to thrive even under exhibition conditions (Hamlaoui et al., 2022). Giraud's project highlights the symbiotic relationship between biological organisms and creative media, and demonstrates the potential of DILS to foster a deeper engagement with living systems.

Similarly, microbiologist and artist Zachary Copfer explores the dynamic potential of living systems in visual communication through his project Bacteriography.[71] Instead of relying on conventional light-sensitive paper, Copfer uses genetically modified bacteria as a living medium. The process involves carefully arranging the bacteria on a petri dish and exposing them to short-wave ultraviolet radiation. This controlled exposure stimulates the bacteria to grow in specific patterns, gradually forming the desired image. Once the bacterial composition matches his artistic vision, Copfer stops the process by cooling the bacteriograph, effectively preserving the microbial artwork in its final state.[72]

Both Giraud's algaegraphy and Copfer's bacteriography exemplify the transformative potential of DILS in creating living art that fuses biological processes with creative expression. These projects underscore the adaptability of microorganisms as interactive agents of visual communication and provide a compelling narrative of how art and science can converge to challenge conventional paradigms of image-making.

In a seminal study, researchers at MIT have engineered a strain of *Escherichia coli* bacteria capable of detecting the RGB (red, green, blue) spectrum of light, effectively giving these microorganisms a form of multicolor vision.[73] This groundbreaking innovation represents a significant advance in the field of DILS, demonstrating how living organisms can be programmed and included in a designed living system to interpret and respond to complex external stimuli (Fernandez-Rodriguez et al., 2017). These genetically modified bacteria are designed to discriminate between RGB wavelengths of light and activate specific genes in response. Each activated gene triggers a distinct biological function, providing a novel method of programming living systems to perform highly specific and dynamic tasks. Achieving this level of precision required the creation of a sophisticated intracellular system within the bacteria, analogous to how a computer chip works. Proteins and enzymes within the cells act as molecular building blocks, processing light signals and triggering precise biological outputs. To illustrate the capabilities of this innovation, the researchers used RGB lighting to guide the bacteria in creating vivid images on culture plates. This included rendering the iconic MIT emblem with remarkable precision, demonstrating the potential of these engineered microorganisms to create intricate, dynamic biological patterns. Such applications highlight the immense versatility of this technology, opening up new possibilities in fields ranging from bioart and data visualization to biosensing and biomanufacturing.

Experimenting with microorganisms not only facilitates the creation of images from scratch, but also enables the modification of existing photographs, producing unexpected and intriguing effects. This approach exemplifies the potential of DILS to transform visual media, offering a dynamic interplay between living organisms and human creativity.

Seung-Hwan Oh, a visionary South Korean artist, explores this potential in his project Impermanence.[74] Oh's innovative technique involves immersing developed photographic portraits in water and introducing a variety of microbes to initiate a transformative process. Over weeks or even years, the microorganisms gradually degrade and consume the silver halide particles in the film's emulsion. This microbial interaction causes shifts in the emulsion and dyes, producing a vivid array of colors and textures. The final portraits are evocative and beautiful, documenting the organic transformation and blurring the boundaries between the original image and its microbially altered state. Oh's work exemplifies how DILS can merge human intent with the autonomous processes of living organisms to create evocative and unpredictable results.

Another pioneering DILS exploration is designer Ori Elisar's Living Language Project,[75] which bridges biology and linguistics to reimagine the evolution of written communication (Figure 3.28). This project uses the unique properties of *Paenibacillus vortex* bacteria to create a living narrative of the evolution of the Hebrew script. Elisar uses a bioink derived from bacterial colonies to inscribe ancient Hebrew letters on Petri dishes. These inscriptions are strategically paired with nutrients in the form of modern Hebrew characters. Over time, the bacteria migrate towards the nutrient sources, morphing and reshaping the colonies to visually emulate the transition from ancient to modern Hebrew letters.

This dynamic process reflects the controlled evolution of language, driven by biological forces and environmental factors such as nutrient availability and ambient temperature (Elisar, 2018). Elisar's work highlights the interplay between human design and the autonomous behavior of living organisms within DILS, demonstrating both the limits of human control and the unpredictable creativity of natural processes. By merging biological research with

Figure 3.28 The Living Language Project (2014) by Ori Elisar – Petri dishes showing the growth patterns of *Paenibacillus vortex* bacteria forming transitions between ancient and modern Hebrew letters.
Credits: Photography by Niva Simon.

creative expression, the Living Language Project transcends the traditional boundaries between art and science. It offers profound insights into the organic evolution of language, highlighting the mutability of alphabets as they adapt to changing cultural contexts. A powerful commentary on the symbiotic relationship between nature and human culture, the project demonstrates how DILS can expand the horizons of linguistic and artistic exploration. Through this blend of biological research and creative expression, Elisar encourages a deeper consideration of the intertwined nature of language development and the living world, highlighting the fluidity and adaptability of alphabets in response to changing cultural landscapes.

The extraordinary ability of microorganisms, especially *flavobacteria*, to emit iridescent structural colors opens innovative possibilities for visual communication. This unique phenomenon is being studied by designers and researchers at TU Delft, in particular in Elvin Karana's Bio Lab.[76] Their work focuses on using these bright colors and patterns to create new forms of interaction and aesthetic expression in designing living artifacts (Kim et al., 2023; Groutars et al., 2022; Risseeuw et al., 2023).

The first concept, called the "living monitor", transforms the flavobacteria's habitat into an interactive display that reflects the user's actions (Figure 3.29). Variations in humidity, influenced by smartwatch data such as daily steps, trigger changes in the shape, texture and color of the bacterial colony. This visual transformation not only mirrors the user's physical movements but also cultivates a deeper relationship between the user and the living display, potentially encouraging healthier behaviors through a sense of empathy.

Taking advantage of flavobacteria's responsiveness to environmental changes, the "living label" concept uses these microorganisms as biological indicators of indoor plant health (Figure 3.30). As soil moisture levels change, the bacteria adapt their appearance to provide a visually intuitive indicator. This bioreactive label enhances the interaction between people and plants, making plant care both more interactive and more informative.

The final concept, "living notes", introduces an innovative approach to communication by using flavobacteria to create messages that change over time (Figure 3.31). Initially invisible, these messages emerge as the bacteria proliferate, only to slowly disappear as the colony expands. This method gives messages an ephemeral, interactive quality, widening the scope for engaging and fluid interactions.

Taken together, these design concepts demonstrate the unique potential of flavobacteria within the framework of DILS, highlighting a broader exploration of the integration of living organisms into design. This approach promotes the creation of interactive, environmentally responsive and living interfaces that enhance our engagement with and understanding of the natural world.

Figure 3.29 a–b **'Flavorium' – the 'Living Monitor' concept demonstrates how Flavobacteria respond to human activity by changing colour and texture according to environmental humidity levels, which are influenced by body movement. The smartwatch data are translated into visual bacterial patterns: (a) represents an inactive lifestyle, showing minimal chromatic variation, while (b) depicts an active lifestyle with more vibrant, dynamic bacterial growth.**

Credits: Groutars and Risseeuw, 2022.

Figure 3.30 'Flavorium' – the 'Living Label' concept.
This illustrates how flavobacteria act as biological indicators of a houseplant's well-being. These bacteria respond to environmental cues, such as soil moisture, temperature and light levels, by altering their colour and texture. This provides a dynamic visual signal of the plant's health.
Credits: Groutars and Risseeuw, 2022.

Figure 3.31 Flavorium – the 'Living Notes' concept uses Flavobacteria as living messengers for playful, time-based communication. The bacteria react to touch, temperature and humidity, leaving temporary colourful traces that gradually evolve and fade. This enables an interactive dialogue between users through living biological patterns.

Credits: Groutars and Risseeuw, 2022.

The annual Agar Art Contest,[77] organized by the American Society of Microbiology (ASM), is a remarkable example of the innovative convergence of science, art and creativity, in line with the principles of DILS. Using microorganisms as artistic media – from naturally pigmented species such as *Serratia marcescens* to genetically engineered strains – this competition transforms standard agar plates into unconventional canvases for dynamic and visually striking works of art.

Each edition of the competition has a unique theme, such as "Microbiology in Space" or "Portrait of Your Favourite Microbiologist", which challenges entrants to combine scientific expertise with artistic expression. This fusion highlights the aesthetic potential of microorganisms and invites the public to explore the intricacies of microbiology through an accessible and engaging medium. Beyond its artistic appeal, the competition has an important educational dimension. It serves as a platform to raise awareness of the immense diversity, behavior and ecological role of microbes, the vast majority of which are invisible to the naked eye. By presenting microorganisms in a creative light, the competition bridges gaps in understanding and appreciation between the scientific community and the general public.

The Agar Art Competition is also in line with the principles of DILS, as it illustrates the interactive potential of living systems within a designed framework. Participants manipulate living organisms to create art, resulting in a dynamic interplay between human intent and microbial behavior. This collaboration between biology and design underscores the transformative potential of integrating living systems into creative practices and echoes the broader possibilities of DILS in other fields such as environmental sustainability, healthcare and materials innovation.

The last case studies of this section, the Nukabot, introduces a novel form of communication between humans and non-human entities, using technology to mediate interactions between users and the microbes involved in fermentation. The Nukabot is an enhanced version of the traditional Japanese wooden fermentation bucket, equipped with chemical sensors and a vocal interface. Unlike its conventional counterpart, the Nukabot offers a dynamic layer of interaction by monitoring the fermentation process and engaging its users through voice commands. This enhanced version of the traditional Japanese *nukadoko* bucket incorporates sensors and a voice interface, translating the invisible microbial processes into a tangible, conversational experience. By personifying the bucket with a blinking eye and a childlike voice, the Nukabot allows users to "talk to" the microbes, making their care an engaging and interactive process.

The voice interface enables users to ask questions like, "How are you doing?" or "How do the pickles taste now?" The Nukabot responds based on real-time chemical data, providing updates about the fermentation process or reminding users to stir the rice bran. These interactions transform what might traditionally be a silent, routine task into a dynamic dialog, fostering a sense of connection between users and the microbes. The playful design, inspired by Japanese *yōkai* folklore, reinforces this relationship by making the Nukabot approachable and relatable. Participants in the study reported that this communication fostered affection and a stronger sense of responsibility. Unlike automated systems, the Nukabot's reliance on human involvement – such as manually stirring the rice bran – encourages users to remain physically and emotionally engaged. The act of speaking to the Nukabot and receiving feedback made users more aware of the microbes as living collaborators. For some, this transformed the Nukabot into a "companion" or "partner", blurring the line between technology and living systems (Chen et al., 2021).

In summary, the integration of living microorganisms into visual and verbal communication represents a paradigm shift in how humans perceive and interact with their environment (Zhou et al., 2023). This confluence of science, art and technology is challenging conventional boundaries and enabling more dynamic and interconnected forms of creative expression. By incorporating biological entities such as bacteria and algae into the visual arts, DILS introduce an innovative framework that merges biology with visual storytelling, fostering a deeper engagement with both the medium and the natural world.

DILS exemplify this transition by inviting a seamless integration of biological processes into design, resulting in living, adaptive and sustainable systems. These systems not only enhance the sensory and aesthetic qualities of visual communication, but also promote a mindset rooted in sustainability and connectedness.

As this interdisciplinary approach continues to evolve, it promises to open up unprecedented possibilities for visual communication, where the boundaries between art, science and the environment are blurred and new paths to creativity are revealed. This marks the beginning of a transformative era – one in which design, empowered by living systems, becomes a medium for fostering deeper connections and creating a sustainable future.

Entertainment

The application of DILS is remarkably broad, extending far beyond the domains originally considered. DILS has made significant advances in areas such as entertainment, with notable implications for fields such as game design and sound design. This ongoing expansion reflects a relentless pursuit of innovative methodologies aimed at fostering dynamic interactions between users, technology and living organisms. In the realm of game design, DILS introduces an unprecedented level of complexity by integrating biological processes directly into gameplay. These systems enable real-time experiences that are dynamically influenced by natural phenomena, offering players a unique and evolving interaction shaped by the behavior of living systems. For example, microbial activity or the growth patterns of organisms could influence game mechanics, creating an immersive and organic link between the virtual and natural worlds.

The versatility of DILS applications continues to highlight its transformative potential across multiple facets of human activity. By embedding biological elements into entertainment technologies, DILS fosters a richer, more symbiotic relationship between people, technology and the natural world. This approach redefines engagement and points to a future where the boundaries between living systems and designed experiences are seamlessly intertwined, fostering innovation that is as sustainable as it is creative.

The EMW Street Bio Team's Biota Beats project[78] is a pioneering example of how DILS can fuse science, art and technology to create new experiences. Using microbiome samples collected from students around the world, the team created a unique 3-minute composition. The process involved growing bacteria from these samples, capturing images of the resulting petri dishes and using a computer algorithm to convert the visual patterns into sound. The resulting composition blended melodies from European oral microbiomes, harmonies from Asian ear bacteria, rhythms from African hand bacteria, percussion from South American scalp microbes and atmospheric sounds from North American nasal bacteria. By collaborating with hip-hop DJs and hosting workshops, Biota Beats extended its reach, enabling participants to create their own music from microbiome data while providing an immersive educational experience that bridged music production and microbiological research.

The BioComputer Music[79] project, a duet between a pianist and the plasmodial slime mold *Physarum polycephalum*, further demonstrates the artistic potential of DILS. Using an analog circuit that incorporates the memristive properties of the slime mold, the system "listens" to and interacts with the pianist's performance, generating musical responses that are transmitted via electromagnets to the piano's strings. This unique collaboration bridges biological systems and musical performance, offering a new perspective on the integration of living organisms with art and technology to create new forms of art and communication (Braund & Miranda, 2016; Miranda & Braund, 2017).

In Algae Opera,[80] created by Michael Burton and Michiko Nitta, an opera singer engages in a remarkable collaboration with living algae, embodying the principles of DILS. This avant-garde performance integrates art, biology and gastronomy, using the singer's breath to stimulate the growth of the algae while simultaneously altering its flavor through the nuanced use of vocal techniques (Stępień, 2021). By tuning the composition and performance to enhance both the biological and sensory properties of the algae, the project demonstrates the dynamic interplay between human creativity and biological systems. The performance offers a multisensory experience, allowing the audience to see, hear and even taste the transformation of the algae, literally consuming the artistic outcome of the performance. This novel approach highlights the potential of DILS to create deeply engaging and interactive experiences that transcend traditional artistic boundaries. It also demonstrates how living organisms can be embedded in creative processes, not just as passive elements, but as active participants who respond to and shape the outcome.

In The Secret Sounds of Spores[81] (Seznec & Hickey, 2013), artists Yann Seznec and Patrick Hickey explore the often ignored beauty of fungi through an innovative art installation that transforms the silent release of mushroom spores into a compelling audiovisual experience. At the heart of the installation is a mushroom enclosed in a glass dome, which isolates it from external air currents to allow precise observation of its spore release (Figure 3.32). A laser beam strategically positioned beneath the mushroom captures the movement of the spores as they fall, producing fleeting flashes of light as the spores pass through the beam. These flashes are

Figure 3.32 The Secret Sounds of Spores (2010–2014) – An interactive art installation by Yann Seznec and Patrick Hickey that translates the silent release of mushroom spores into sound and light.
Credits: Yann Seznec.

recorded in real time by a digital video camera, and the data is processed by a computer system to control a series of bespoke musical instruments suspended above the installation. The result is an emergent and dynamic composition in which the mushroom itself becomes the conductor, orchestrating a unique blend of visual and auditory elements.

This project not only highlights the aesthetic and ecological importance of fungi, but also illustrates the interactive potential of DILS. By integrating biological processes into the creative medium, The Secret Sounds of Spores transforms a natural phenomenon into an engaging sensory experience. The interplay between the falling spores, laser light and responsive musical instruments underlines the importance of real-time interactions between living systems and designed environments, a hallmark of DILS.

The installation also encourages viewers to rethink their relationship with nature by making the invisible visible and the inaudible audible. It challenges traditional notions of static art by introducing biological agency, offering an ever-changing experience that evolves with the mushroom's lifecycle.

Nicola Burggraf's Interference project[82] (2014) is a groundbreaking exploration at the intersection of biology, art and technology, using bioluminescent algae to make audible frequencies visible. In this project, dense clusters of bioluminescent algae are suspended in a saline solution where they respond to sound waves with remarkable precision. When exposed to certain frequencies, the algae emit bursts of bluish light, effectively translating auditory vibrations into a visually arresting display. The natural luminescent properties of the algae serve as a medium for this transformation, allowing viewers to perceive sound waves as tangible, fleeting patterns of light. By exploring the biological responsiveness of algae, Interference invites audiences to engage with the interplay of sound and light in a way that is both scientifically intriguing and artistically engaging. This real-time interaction between biological systems and acoustic stimuli demonstrates the unique potential of DILS to create immersive environments that dissolve traditional boundaries between natural phenomena and designed experiences.

Again, a meaningful example of DILS in action highlights the transformative power of biological processes in expanding the possibilities of artistic expression and interactive design. It redefines how living systems can be used to create novel sensory experiences, further illustrating the interdisciplinary potential of this emerging field.

Raphael Kim's Mould Rush project[83] redefines the boundaries of game design by integrating DILS into digital gaming platforms. This innovative online multiplayer strategy game seamlessly blends tangible biological processes with digital interactions, offering a unique gaming experience centered on the real-time growth and behavior of living microorganisms. Unlike traditional gaming approaches, where fast-paced action often dominates, Mould Rush uses the deliberate pace of microbial development as an enriching feature rather than a limitation (Kim et al., 2018, 2019, 2020).

In addition to providing entertainment, Mould Rush is intended to be an educational platform, introducing players to basic microbiological concepts such as cell culture, colony morphology and the behavioral dynamics of microbial colonies. The strategic elements of the game, coupled with the deliberate pace of microbial growth, are designed to develop players' skills of memory, observation and patience, promoting a unique blend of scientific learning and perseverance.

The project highlights the potential of DILS to create dynamic interactions between players and living systems. Within the game, the growth and morphology of microbial colonies influence gameplay, adding a layer of complexity shaped by biological processes. This biotic dimension provides players with a novel, evolving interaction that fosters a deeper engagement with natural phenomena. By immersing players in the slow but intricate dynamics of microbial development, Mould Rush transforms the traditionally passive observation of biology into an interactive and participatory experience.

Beyond its entertainment value, Mould Rush serves as an educational platform, introducing players to basic microbiological concepts such as cell culture, colony morphology and the behavioral dynamics of microorganisms. By interacting with living systems within the game, players gain insight into the underlying biological processes that govern microbial growth. This hands-on engagement cultivates memory, observation and patience, fostering a deeper understanding and appreciation of the complexities of microbial life. This design approach emphasizes the versatility of DILS in reshaping how people engage with technology and the natural world, fostering a symbiotic relationship that promotes both scientific literacy and creative exploration.

These examples demonstrate the enormous potential of DILS to revolutionize traditional entertainment and educational contexts. By connecting living organisms with human creativity and technological innovation, DILS offer a future rich in deep and meaningful exchanges between biological systems and technological advances. These pioneering projects deepen the understanding and appreciation of biology, and also explore new possibilities for interactive experiences. In doing so, DILS are redefining entertainment as a medium that fosters dynamic interactions between people, technology and the living world, encouraging novel, immersive and ecologically attuned relationships.

CONCLUSIONS

The integration of DILS into various domains – from entertainment to medicine and beyond – represents a transformative paradigm shift in design for interaction. By embedding living organisms into interactive systems, DILS offer pioneering alternatives for sustainable and regenerative practices. These systems minimize the use of raw materials, operate with exceptionally low energy requirements and fit seamlessly with circular design principles, providing a model for innovation based on environmental responsibility. In addition to revolutionizing the functionality of interactive systems, DILS are reshaping the language of interaction and redefining human relationships with the more-than-human world. By fostering symbiotic engagements rather than exploitative dynamics, these systems promote a reconnection with nature rooted in collaboration and mutual benefit. Moving beyond anthropocentric frameworks, DILS invite humans to participate as co-creators within ecological systems, cultivating harmony and equitable coexistence between humans and non-human entities.

The range of applications explored in this chapter demonstrates the versatility and transformative potential of DILS. These systems invite a rethinking of interaction design, shifting users from passive to active participants in their relationship with living artifacts, fostering a sense of responsibility and engagement.

The case studies collected reflect also a design philosophy rooted in ecological principles, emphasizing mutualism, co-evolution and sustainable coexistence. These projects transcend human-centered limitations, exploring temporal, scalar and aesthetic dimensions that foster connections between humans and the more-than-human world (Karana et al., 2023). While some projects have achieved practical application, others remain speculative, outlining potential futures and promoting interdisciplinary research that bridges art, design and science.

Nevertheless, the ethical discourse around DILS is still in its formative stages. Much work remains to be done to understand the implications of integrating living organisms into design systems, including the ethical considerations of collaborating with non-human entities.

Ultimately, DILS represent a transformative step forward, challenging conventional norms of interaction design while aligning with sustainable and regenerative philosophies. Through these groundbreaking projects, DILS is leading the way to a future where innovation and ethical responsibility coexist, shaping a more sustainable, equitable and connected world.

NOTES

1 The Human Genome Project was a large, well-organized and highly collaborative international effort that generated the first sequence of the human genome and that of several additional well-studied organisms. Carried out from 1990 to 2003, it was one of the most ambitious and

important scientific endeavors in human history: https://www.genome.gov/about-genomics/educational-resources/fact-sheets/human-genome-project#:~:text=The%20Human%20Genome%20Project%20was,90%25%20of%20the%20human%20genome.

2 Alive and Kicking: Designing with Living Materials by Elvin Karana: https://www.tudelft.nl/en/stories/articles/alive-and-kicking-designing-with-living-materials Living Materials by Franklin and Till: https://www.franklintill.com/journal/living-materials Hybrid Living Materials by Neri Oxman: https://oxman.com/projects/hybrid-living-materials

3 Hiroshi Ishii is Jerome B. Wiesner Professor of Media Arts and Sciences, associate director of MIT Media Laboratory and director of Tangible Media Group at Massachusetts Institute of Technology. https://tangible.media.mit.edu/person/hiroshi-ishii/

4 Some examples are Meredith Patterson's *Biopunk Manifesto* (*A Biopunk Manifesto* by Meredith L. Patterson (2010): http://dpya.org/wiki/index.php/2010_-_A_Biopunk_Manifesto_-_Meredith_L._Patterson) and the launch of the DIYbio.org forum in 2008. This forum has become a vital institution for DIY biologists, serving as a hub for collaboration and innovation. It is in line with the ethos of making biology an accessible activity for citizen scientists, amateur biologists and DIY biological engineers who value openness and safety.

5 Building with Biology to Benefit All People and the Planet: https://biobricks.org/

6 OpenWetWare is an effort to promote the sharing of information, know-how and wisdom among researchers and groups who are working in biology and biological engineering: https://openwetware.org/wiki/Main_Page

7 Your open-source PCR Thermocycler: https://openpcr.org/

8 BigBIO (BigScience Biomedical) is an open library of biomedical dataloaders built using Huggingface's (🤗) datasets library for data-centric machine learning: https://hal.science/hal-03723519/

9 Find out more about Professor Derek Lovley's research activities: https://www.micro.umass.edu/faculty-and-research/derek-lovley

10 "Developing 'green' electronics: Team finds microbe from the Potomac yields better electronic material", 2017: https://phys.org/news/2017-01-green-electronics-team-microbe-potomac.html https://www.umass.edu/news/article/expanding-potential-functions-electrically

11 Bioelectronics Smart Surfaces – BIOSS: https://www.itodys.univ-paris-diderot.fr/fr/recherche/departement-1/biosensors-and-smart-surfaces-bioss-d11

12 "Living sensors at your fingertips. Cell-infused gloves and bandages light up when in contact with certain chemicals". Available at: https://news.mit.edu/2017/living-sensors-your-fingertips-0215

13 Harpreet Sareen is a designer, researcher and artist creating mediated digital interactions through the living world, with growable electronics, organic robots and bionic materials. For more information: https://harpreetsareen.com/

14 "Cyborg botany: Augmented plants as sensors, displays, and actuators": https://www.media.mit.edu/projects/cyborg-botany/overview/ https://www.youtube.com/watch?v=3fsE_c__zV4 "MIT researchers

engineer 'cyborg' plants for motion-tracking and sending notifications": https://www.dezeen.com/2019/06/17/mit-researchers-engineer-cyborg-plants/

15 "Botanicus Interacticus: Interactive Plant Technology": https://satomunehiko.com/works/botanicus-interacticus/

16 "How Glowing Bacteria in the Dirt May One Day Save Lives" (2021): https://www.nytimes.com/2021/06/21/science/landmines-glowing-bacteria.html

17 "Bacteria as a Digital Mediator", RCA2020, Royal College of Art 2020; https://akvilete.cargo.site/Bacteria-as-a-Digital-Mediator

18 Akvile Terminaite, Experience Designer: https://2020.rca.ac.uk/students/akvile-terminaite

19 "Ingestible 'bacteria on a chip' could help diagnose disease": https://news.mit.edu/2018/ingestible-bacteria-on-a-chip-help-diagnose-disease-0524 The ingestible bacterial-electronic sensor: https://www.youtube.com/watch?v=Jg2aD4mc-bw

20 "The Ingestible Bacterial-Electronic Sensor": https://engineering.mit.edu/engage/engineering-in-action/the-ingestible-bacterial-electronic-sensor/

21 iGEM 2009 Jamboree, October 31 to November 2, 2009 – Massachusetts Institute of Technology https://2009.igem.org/Team:Cambridge; https://2009.igem.org/Main_Page

22 "E. chromi: Living Colour from Bacteria": https://www.daisyginsberg.com/work/echromi-living-colour-from-bacteria

23 Future Flora is a harvesting kit designed for women to treat and prevent vaginal infections: https://gitomasello.com/future-flora You can also find more information here: https://www.dezeen.com/2016/07/05/giulia-tomasello-future-flora-home-grown-sanitary-pad-central-saint-martins/

24 Report of the World Commission on Environment and Development: Our Common Future: https://sustainabledevelopment.un.org/content/documents/5987our-common-future.pdf

25 "Clean vs. dirty energy: the fight for sustainability": https://nchschant.com/21326/investigative/clean-vs-dirty-energy-the-fight-for-sustainability/

26 GeoMerce Project: https://www.giovanniinnella.com/Design-Practice/geomerce; http://www.gionatagatto.com/Geomerce_Concept.html

27 The Institute for Advanced Architecture of Catalonia. Located in Barcelona and operating across the world, AAG investigates emerging technologies of computation and digital manufacturing combined with scientific advancements from the fields of data science, biology, material science, ecology and robotics: https://iaac.net/

28 Baiobabes: https://www.biobabes.co.uk/biocatalyticcell Biophoto-voltaic (BPV) Algae Sensor: https://www.iaacblog.com/programs/biocatalytic-cell/

29 "Wallpaper bio-solar panel developed by researchers": https://www.imperial.ac.uk/news/182936/wallpaper-bio-solar-panel-developed-researchers/

30 "Producing 'green' energy – literally – from living plant 'bio-solar cells'": https://www.technion.ac.il/en/2023/01/energy-plant-bio-solar-cells/

31 Electric life project's description: https://www.teresavandongen.com/Electric-Life
32 Designer's personal web site: https://www.teresavandongen.com/
33 Living Light description: https://livinglight.info/living-light/
34 Living Light is an off-grid lamp powered by photosynthesis: https://www.dezeen.com/2017/11/15/living-light-ermi-van-oers-microbial-energy-photosynthesis-lighing-lamp-good-design-bad-world-dutch-design-week/
35 ecoLogicStudio is an architecture and design innovation firm specializing in biotechnology for the built environment: https://www.ecologicstudio.com/about
36 Habitat One envisions the architecture of the carbon neutral city: https://www.ecologicstudio.com/projects/habitat-one-the-architecture-of-carbon-neutral-city
37 "Tree one: an artificial tree for a generation of a carbon neutral world": https://motorstudio.hyundai.com/busan/cotn/exhb/habitatOne.do?strgCd=04&lang=en
38 H.O.R.T.U.S. XL Astaxanthin.g, a living bio-structure demonstrating the possibility of coexistence of nature and city: https://motorstudio.hyundai.com/busan/cotn/exhb/habitatOne.do?strgCd=04&lang=en
39 *Living Things* is a collaborative project by David Benjamin, Soo-in Yang and others, exploring the integration of living microalgae into built environments. Retrieved from: http://www.ethanfrier.com/living-things; http://www.douenias.design/living-things-1
40 Exhale – The Bionic Chandelier is a living sculpture by Julian Melchiorri that uses microalgae to purify air and produce oxygen. Retrieved from: https://www.julianmelchiorri.com/Bionic-Chandelier; https://www.youtube.com/watch?v=Lxd43iH-CX0
41 Glowpolis is a speculative project by Glowee envisioning a city illuminated by bioluminescent organisms. Retrieved March, 2025, from https://en.glowee.com/glowpolis
42 GlowEvent is an immersive installation by Glowee that showcases bioluminescence as a sustainable lighting solution. Retrieved March, 2025, from https://www.glowee.com/glowevent
43 Biolumia is a bioluminescent lighting installation developed by Glowee to explore sustainable urban illumination. Retrieved March, 2025, from https://www.glowee.com/biolumia
44 Growing Light by Tim van Dortmont (2021) is an interactive installation using algae-based bioluminescence to visualize the energy transition. Retrieved March, 2025, from https://www.highlightdelft.nl/program/in2021/growing-light/
45 Ambio, bacterial lamp: https://www.teresavandongen.com/Ambio
46 "Discover the subtle soothing luminescence of the Firefly Petunia": https://light.bio/
47 Biology by Design: https://www.ginkgobioworks.com/
48 "What if we could enable others to see bacteria as a carrier of meaning, rather than a carrier of disease?": https://www.janklingler.com/bacteria-in-a-new-light

49 "Jan Klingler's Bacteria lamps are 'modern fossils' as well as lights": https://www.dezeen.com/2019/02/17/bacteria-lamps-jan-klingler-design/
50 "Investigation into Collaboration Between Fashion Design Research and Biology": https://research-biennale.rca.ac.uk/projects/vital-assemblages-a-fashion-led-research-investigation-into-collaboration-between-fashion-design-research-and-biology
51 Bioluminescence exhibit featured on Design Exchange: https://www.engbio.cam.ac.uk/news/bioluminescence_Design_Exchange
52 Lucid Life is a speculative design project by Biocrafted that explores the potential of engineered bioluminescent organisms as living light sources. Retrieved March, 2025, from https://biocrafted.com/lucid-life
53 Center for Microbial Ecology and Technology: https://www.cmet.ugent.be/users/dr-ir-chris-callewaert
54 bioLogic by Lining Yao, Wen Wang, Guanyun Wang, Helene Steiner, Chin-Yi Cheng, Jifei Ou, Oksana Anilionyte, Hiroshi Ishii: https://tangible.media.mit.edu/project/biologic/
55 Biogarmentry: https://www.materialincubator.com/biogarmentry
56 BioRealize: Microbial Design Studio Prototype: https://www.jayhardman.com/project-3
57 "Puma and MIT Design Lab envision a future of self-adapting, performance-enhancing sportswear": https://www.dezeen.com/2018/05/24/mit-media-lab-puma-future-sportswear-design/
58 "Introducing MIT Design Lab x PUMA Collaboration: Adaptive Dynamics": https://www.youtube.com/watch?v=EWK5mlv-j9g The term "diegetic prototypes" was introduced by film scholar David Kirby in 2010. It refers to technologies that exist only within the fictional world of a narrative, known as the diegesis, but function as fully functional objects within that world. These diegetic prototypes are used in films to communicate the necessity, feasibility and positive aspects of future technologies to a wide audience through dialog, plot justification, character interactions and narrative structure. Diegetic prototypes result from a combination of prototyping and storytelling, emphasizing the creation of objects (artifacts of a story world) to tell stories effectively in the context of design fiction. Kirby, D. (2010). "The future is now: Diegetic prototypes and the role of popular films in generating real-world technological development". *Social Studies of Science*, 40(1), 41–70. https://doi.org/10.1177/0306312709338325 Li, L. (2023). "Diegetic Prototypes in the Design Fiction Film Her: A Posthumanist Interpretation". *Journal of Futures Studies*, 27(3), 63. https://doi.org/10.6531/JFS.202303_27(3).0004.
59 For more details about this project, you can visit Shamees Aden's website at http://shameesaden.com/
60 "A biosensor to detect & sync human emotions in collaboration with Ka Hei Suen, Christian Felsner and Takeuchi Lab", 2017: https://www.charlottefuret.com/chemical-synchronization
61 A living, digital display made of bioluminescent bacteria: https://www.charlottefuret.com/chameleon
62 For more information: https://www.smartcitiesdive.com/ex/sustainablecitiescollective/bioluminescent-billboards/33364/

63 Hunter Cole, artist and scientist, formerly known as Hunter O'Reilly, is recognized internationally for producing work that is inspired by science, but lives as art: https://www.huntercole.org/
64 "Hunter Cole – Her Own DNA – Living Drawings Created with Bioluminescent Bacteria with Protein Music": https://www.youtube.com/watch?v=2KYDTYcDtKQ
65 Contagion – Bacteria Billboard: https://www.youtube.com/watch?v=LppK4ZtsDdM
66 Patrick Hickey is a mycologist working with bioluminscent fungi: https://www.mediamatic.net/en/page/70264/patrick-hickey
67 The Elecrtic Factory: https://theelectricfactory.com/
68 "The Electric Factory Grows Out-Of-Home Bacteria Ads to Encourage Hand-Washing": https://www.lbbonline.com/news/the-electric-factory-grows-out-of-home-bacteria-ads-to-encourage-hand-washing
69 For more information: https://www.liagiraud.com/
70 "The Unmasking of Image": https://jeudepaume.org/en/palm/algaegraphs-unmasking-the-image-essay-by-lia-giraud-palm-jeu-de-paume-2/
71 For more information: http://www.sciencetothepowerofart.com/bacteriography
72 "Bacteriography – An Explanation", by Zachary Copfer: https://www.youtube.com/watch?v=kvMBVJozRI0
73 "Bacteria with multicolor vision", 2017: https://news.mit.edu/2017/bacteria-multicolor-vision-0526
74 For more information: https://www.seunghwan-oh.com/impermanence
75 The Living Language is an interdisciplinary BioDesign project by artist and designer Oriel IsaR, exploring communication through living systems. Retrieved March, 2025, from https://orielisar.com/
76 The work of Prof Elvin Karana, in addition to her many seminal publications, can also be found in the following references: https://www.tudelft.nl/io/over-io/personen/karana-e; https://www.materialincubator.com/projects; https://materialsexperiencelab.com/
77 Established in 1899, American Society of Microbiology is the home for microbial scientists from around the globe to connect, learn, discover and prepare for the future. They organize the ASM Agar Art Contest: https://asm.org/Events/ASM-Agar-Art-Contest/Home
78 "A microbial record player, translating bacteria to sound": http://biotabeats.org/ "Biota Beats: What if you could make music from your microbiota?": https://ani-liu.com/biota-beats
79 "This Musician Plays Piano Duets with a Microorganism": https://www.labiotech.eu/more-news/eduardo-miranda-music-biocomputing/ Biocomputer Music is an experimental one-piano duet for a pianist and a biocomputer: https://soundcloud.com/ed_miranda/biocomputer-music
80 The Algae Opera, 2012: https://www.burtonnitta.co.uk/AlgaeOpera.html
81 For more information: https://www.yannseznec.com/works/spores/
82 Bioluminescent Installation, 2012: http://www.nicolaburggraf.com/interference.html
83 "Playing Computer Games with ... Bacteria?": https://medium.com/@rphlkim/playing-computer-games-with-bacteria-739b57488c66

REFERENCES

Adiloglu, F. (2011). Visual communication: Design studio education through working the process. *Procedia - Social and Behavioral Sciences*, *28*, 982–991. https://doi.org/10.1016/j.sbspro.2011.11.182

Aghighi, R. (2019). What if our clothes were alive and photosynthesized? *APRIA Journal*, 121–126. Retrieved March, 2025, from http://apria.artez.nl/what-if-our-clothes-were-alive-and-photosynthesized/

Amobonye, A., Lalung, J., Awasthi, M. K., & Pillai, S. (2023). Fungal mycelium as leather alternative: A sustainable biogenic material for the fashion industry. *Sustainable Materials and Technologies*, *38*, e00724. https://doi.org/10.1016/j.susmat.2023.e00724

Antonacci, A., & Scognamiglio, V. (2019). Biotechnological advances in the design of algae-based biosensors. *Trends in Biotechnology*, *38*(3), 334–347. https://doi.org/10.1016/j.tibtech.2019.10.005

Ahmad, A. (2020). *Fashion's bio-fabrication revolution*. New Degree Press.

Arkin, A. P., & Endy, D. (1999). A standard parts list for biological circuitry (DARPA white paper). DARPA. Retrieved from https://dspace.mit.edu/bitstream/handle/1721.1/29794/Arkin.Endy.DARPA.pdf

Ast, J. C., Cleenwerck, I., Engelbeen, K., Urbanczyk, H., Thompson, F. L., De Vos, P., & Dunlap, P. V. (2007). Photobacterium kishitanii sp. nov., a luminous marine bacterium symbiotic with deep-sea fishes. *International Journal of Systematic and Evolutionary Microbiology*, *57*(Pt 9), 2073–2078. https://doi.org/10.1099/ijs.0.65153-0

Atkinson, J. T., Chavez, M. S., Niman, C. M., & El-Naggar, M. Y. (2022). Living electronics: A catalogue of engineered living electronic components. *Microbial Biotechnology*, *16*, 507–533. https://doi.org/10.1111/1751-7915.14171

Aydemir, C., & Özsoy, S. (2020). Environmental impact of printing inks and printing process. Retrieved March, 2025, from https://www.researchgate.net/publication/346549027_Environmental_impact_of_printing_inks_and_printing_process

Aziz, M. S., & El Sherif, A. Y. (2016). Biomimicry as an approach for bio-inspired structure with the aid of computation. *Alexandria Engineering Journal*, *55*(1), 707–714. https://doi.org/10.1016/j.aej.2015.10.015

Barati, B., Karana, E., Pont, S., & Van Dortmont, T. (2021). Living light interfaces: An exploration of bioluminescence aesthetics. In *DIS 21: Proceedings of the 2021 ACM designing interactive systems conference conference*, June 2021 (pp. 1215–1229). https://doi.org/10.1145/3461778.3462038

Batmunkh, A. (2022). Carbon footprint of the most popular social media platforms. *Sustainability*, *14*(4), 2195. https://doi.org/10.3390/su14042195.

Belkhir, L., & Elmeligi, A. (2018). Assessing ICT global emissions footprint: Trends to 2040 & recommendations. *Journal of Cleaner Production, 177*, 448–463. https://doi.org/10.1016/j.jclepro.2017.12.239

Belkin, S., Yagur-Kroll, S., Kabessa, Y., Korouma, V., Septon, T., Nussinovitch, A., Novodvorets, A., Fishilevich, E., & Shoham, Y. (2017). Remote detection of buried landmines using a bacterial sensor. *Nature Biotechnologies, 35*, 308–310. https://doi.org/10.1038/nbt.3791ù

Benyus, J. M. (2002). *Biomimicry: Innovation inspired by nature* (2nd ed.). Harper Perennial.

Bilgen, S. (2014). Structure and environmental impact of global energy consumption. *Renewable and Sustainable Energy Reviews, 38*, 890–902. https://doi.org/10.1016/j.rser.2014.07.004

Bond, D. R., & Lovley, D. R. (2003). Electricity production by Geobacter sulfurreducens attached to electrodes. *Applied and Environmental Microbiology, 69*(3), 1548–1555. https://doi.org/10.1128/AEM.69.3.1548-1555.2003

Braund, E., & Miranda, E. R. (2016). BioComputer music: Generating musical responses with physarum polycephalum-based memristors. In R. Kronland-Martinet, M. Aramaki, & S. Ystad (Eds.), *Music, mind, and embodiment (*CMMR 2015, Lecture Notes in Computer Science, Vol. 9617). Springer. https://doi.org/10.1007/978-3-319-46282-0_26

Brayner, R., Couté, A., Livage, J., Perrette, C., & Sicard, C. (2011). Micro-algal biosensors. *Analytical and Bioanalytical Chemistry, 401*, 581–597. https://doi.org/10.1007/s00216-011-5107-z

Broadhead, R., Craeye, L., & Callewaert, C. (2021). The future of functional clothing for an improved skin and textile microbiome relationship. *Microorganisms*, 9(6), 1192. https://doi.org/10.3390/microorganisms9061192

Broadhead, R., Craeye, L., & Callewaert, C. (2022). Odour control or inhibition using antimicrobial finishing. In G. Thilagavathi & R. Rathinamoorthy (Eds.), *Odour in textiles: Mechanisms of generation and control.* Psgtech, CRC Press.

Brooks, R. R., & Robinson, B. H. (1998). The potential use of hyperaccumulators and other plants for phytomining. In R. R. Brooks (Ed.), *Plants that hyperaccumulate heavy metals: Their role in phytoremediation, microbiology, archaeology, mineral exploration and phytomining* (pp. 327–356). CAB International.

Brundtland, G. H. (1987). *Our common future: Report of the world commission on environment and development.* UN-Dokument A/42/427. Retrieved March, 2025, from http://www.un-documents.net/ocf-ov.htm

Burggraf, N. (2014). Bioluminescence: Toward design with living light. In M. Kretzer & L. Hovestadt (Eds.), *ALIVE: Advancements in adaptive architecture*. Birkhäuser. https://doi.org/10.1515/9783990436684.82

Camere, S., & Karana, E. (2018). Fabricating materials from living organisms: An emerging design practice. *Journal of Cleaner Production*, *186*, 570–584.

Campbell, A. K. (2012). Darwin shines light on the evolution of bioluminescence. *Luminescence*, *27*(6), ii–ii. https://doi.org/10.1002/bio.2439

Chen, D., Seong, Y., Ogura, H., Mitani, Y., Sekiya, N., & Moriya, K. (2021). Nukabot: Design of care for humanmicrobe relationships. In *Proceedings of the conference on human factors in computing systems* (Article No. 291). ACM. https://doi.org/10.1145/3411763.3451605

Cheng, Y. Y., Chen, Z., Cao, X., Liu, H., Sun, Y., Zhang, W., Wang, Y., & Zhao, Y. (2023). Programming bacteria for multiplexed DNA detection. *Nature Communications*, *14*, 2001. https://doi.org/10.1038/s41467-023-37582-x

Cobbing, M., & Vicaire, Y. (2016). *Timeout for fast fashion*. Greenpeace. Retrieved March, 2025, from http://www.greenpeace.org/international/Global/international/briefings/toxics/2016/Fact-Sheet-Timeout-for-fast-fashion.pdf

Cole, H. (2018). Living Light: Photographs by the Light of Bioluminescent Bacteria [Exhibition catalogue]. Loyola University New Orleans.

Collet, C. (2021). Biotextiles: Making textiles in a context of climate and biodiversity emergency. In O. Pedgley, V. Rognoli, & E. Karana (Eds.), *Material experience 2: Expanding territories of materials and design* (pp. 207–224). Butterworth-Heinemann.

Cross, R. (2017, April 11). Glowing bacteria detect buried landmines. *Science*. https://doi.org/10.1126/science.aal1028

Danino, T., Mondragón-Palomino, O., Tsimring, L., & Hasty, J. (2010). A synchronized quorum of genetic clocks. *Nature*, *426*, 326–330.

Delgado, A. (2013). DIYbio: Making things and making futures. *Futures*, *48*, 65–73. https://doi.org/10.1016/j.futures.2013.02.004

Dincer, I. (1999). Environmental impacts of energy. *Energy Policy*, *27*(14), 845–854. https://doi.org/10.1016/S0301-4215(99)00068-3

Dong, X., Wu, W., Pan, P., & Zhang, X.-Z. (2023). Engineered living materials for advanced diseases therapy. *Advanced Materials*, *35*(29), 2304963. https://doi.org/10.1002/adma.202304963

Dybas, C. L. (2019). A new wave: Science and art meet in the sea. *Oceanography*, *32*(1), 10–11. https://doi.org/10.5670/oceanog.2019.104 – Retrieved from Loyola University.

Dyson, F. (2007). Our biotech future. *The New York Review of Books*, *54*(12), 4–8. Retrieved August, 2023, from https://hahana.soest.hawaii.edu/cmoreserver/summercourse/2012/documents/chisholm_07-02-12/Dyson-Our_Biotech_Future_by_Freeman_Dyson.pdf

Dunlap, P. V., & Urbanczyk, H. (2013). Luminous bacteria. In E. F. DeLong, E. Stackebrandt, S. Lory, & F. Thompson (Eds.), *The Prokaryotes: Prokaryotic physiology and biochemistry* (pp. 495–528). Springer Nature. https://doi.org/10.1007/978-3-642-30141-4_75

Dunn, K. E. (2020). The emerging science of electrosynbionics. *Bioinspiration and Biomimetics*, *15*, 033001. https://doi.org/10.1088/1748-3190/ab654f

Eckelman, M. J., & Sherman, J. (2016). Environmental impacts of the U.S. health care system and effects on public health. *PLoS One*, *11*(6), e0157014. https://doi.org/10.1371/journal.pone.0157014

Elisar, O. (2018, May 15–17). The Living Language Project. In Proceedings of the Politics of the Machines – Art and After Conference, Aalborg University, Copenhagen, Denmark. Retrieved March 2025, from https://www.scienceopen.com/hosted-document?doi=10.14236/ewic/EVAC18.42

Endy, D. (2005). Foundations for engineering biology. *Nature*, *438*(7067), 449–453. https://doi.org/10.1038/nature04342

Ertürkan, H., Karana, E., & Mugge, R. (2022, June 25–July 3). Is this alive? Towards a vocabulary for understanding and communicating living material experiences. In D. Lockton, S. Lenzi, P. Hekkert, A. Oak, J. Sádaba, & P. Lloyd (Eds.), *DRS2022: Bilbao*. Bilbao, Spain. https://doi.org/10.21606/drs.2022.796

Fernandez-Rodriguez, J., Moser, F., Song, M., & Voigt, C. A. (2017). Engineering RGB color vision into Escherichia coli. *Nature Chemical Biology*, *13*, 706–708. https://doi.org/10.1038/nchembio.2390

Garner, K. L. (2021). Principles of synthetic biology. *Essays in Biochemistry*, *65*(5), 791–811. https://doi.org/10.1042/EBC20200059

Ginsberg, A. D., & Chieza, N. (2018). Other biological futures. *Journal of Design and Science*, *2018*, 6868b5.

Goers, L., Kylilis N., Tomazou, M., Wen, K. Y., Freemont, P., & Polizzi, K. (2013). Engineering microbial biosensors. In C. Harwood & A. Wipat (Eds.), *Methods in microbiology* (Vol. 40, pp. 119–156). Academic Press. https://doi.org/10.1016/B978-0-12-417029-2.00005-4

Gomez-Marquez, J., & Hamad-Schifferli, K. (2019). Distributed biological foundries for global health. *Advanced Healthcare Materials*, *8*(18), 1900184. https://doi.org/10.1002/adhm.201900184

Gough, P., Yoo, S., Tomitsch, M., & Ahmadpour, N. (2021). Applying bioaffordances through an inquiry-based model: A literature review of interactive biodesign. *International Journal of Human–Computer Interaction*, *37*(17), 1583–1597. https://doi.org/10.1080/10447318.2021.1898846

Gough, P., Ahmadpour, N., Hepburn, L.-A., Pschetz, L., Catts, O., Cooper, C., & Ramirez-Figueroa, C. (2020). The nature of biodesigned systems: Directions for HCI. In *Companion Publication of the 2020 ACM Designing Interactive Systems Conference (DIS' 20 Companion)*

(pp. 389–392). Association for Computing Machinery. https://doi.org/10.1145/3393914.3395908

Gershenfeld, N. (2005). *Fab: The coming revolution on your desktop—from personal computers to personal fabrication*. Basic Books.

Groutars, E., Risseeuw, C., Ingham, C., Hamidjaja, R., Elkhuizen, W., Pont, S., & Karana, E. (2022). Flavorium: An exploration of flavobacteria's living aesthetics for living color interfaces. In *Proceedings of the 2022 CHI Conference on Human Factors in Computing Systems (CHI '22)* (Article 99, 1–19). Association for Computing Machinery. https://doi.org/10.1145/3491102.3517713

Hamlaoui, S., Yéprémian, C., Duval, C., Marie, B., Djédiat, C., Piquet, B., Bernard, C., & Duperron, S. (2022). The culture collection of cyanobacteria and microalgae at the French National Museum of natural history: A century old but still alive and kicking! *Cryptogamie, Algologie*, *43*(3), 41–83. https://doi.org/10.5252/cryptogamie-algologie2022v43a3

Hobson, B. (2015). Teresa van Dongen's bioluminescent lamp is "a new way to create light." Retrieved August, 2023, from https://www.dezeen.com/2015/01/02/movie-teresa-van-dongen-ambiobioluminescent-bacteria-lamp-video-interview/

Ishii, H. (2008). Tangible bits: Beyond pixels. In Proceedings of the *2nd international conference on Tangible and Embedded Interaction* (TEI '08). ACM. http://dx.doi.org/10.1145/1347390.1347392

Ishii, H., Lakatos, D., Bonanni, L., & Labrune, J.-B. (2012). Radical atoms: Beyond tangible bits, toward transformable materials. *Interactions*, *19*(1), 38–51. Retrieved March, 2025, from https://web.media.mit.edu/~ishii/RadicalAtoms_ACM_Interactios.pdf

Ishii, H., & Ullmer, B. (1997). Tangible bits: Towards seamless interfaces between people, bits and atoms. In *Proceedings of the ACM SIGCHI conference on human factors in computing systems* (CHI '97) (pp. 234–241). ACM. http://dx.doi.org/10.1145/258549.258715

Jabr, F. (2016). The secret history of bioluminescence. *Hakai Magazine*. Retrieved March, 2025, from https://hakaimagazine.com/features/secret-history-bioluminescence/

Jacometti, V. (2019). Circular economy and waste in the fashion industry. *Laws*, *8*(4), 27. https://doi.org/10.3390/laws8040027

Jeremijenko, N. (2012). The environmental medic: XClinic. In L. Nelischer (Ed.), *Future practice: Conversations from the edge of architecture* (pp. 254–266). Taylor and Francis.

Jiménez-Rodríguez, M. G., Silva-Lance, F., Parra-Arroyo, L., Medina-Salazar, D. A., Játiva, C., Pacheco, J., & García-Ruiz, C. (2022). Biosensors for the detection of disease outbreaks through wastewater-based epidemiology. *Trends in Analytical Chemistry*, *155*, 116585. https://doi.org/10.1016/j.trac.2022.116585

Jaiswal, K. K., Chowdhury, C. R., Yadav, D., Verma, R., Dutta, S., Jaiswal, K. S., SangmeshB, & Karuppasamy, K. S. K. (2022). Renewable and sustainable clean energy development and impact on social, economic, and environmental health. *Energy Nexus*, *7*, 100118. https://doi.org/10.1016/j.nexus.2022.100118

Johns, P. (2021). How is bioluminescence employed by humans & in nature? *Journal of Plant Biology and Agriculture Sciences*, *5*(5), Article 224. Retrieved March, 2025, from https://www.imedpub.com/abstract/how-is-bioluminescence-employed-by-humans--in-nature-38640.html

Kapsali, V. (2021). *Biomimetics for designers: Applying nature's processes & materials in the real world* (1st ed.). Thames & Hudson.

Karana, E., Barati, B., & Giaccardi, E. (2020). Living artefacts: Conceptualizing livingness as a material quality in everyday artefacts. *International Journal of Design*, *14*(3), 37–53.

Karana, E., McQuillan, H., Rognoli, V., & Giaccardi, E. (2023). Living artefacts for regenerative ecologies. *Research Directions: Biotechnology Design*, *1*, e16. https://doi.org/10.1017/btd.2023.10

Kahlke, T., & Umbers, K. D. (2016). Bioluminescence. *Current Biology*, *26*(8), 313–314. https://doi.org/10.1016/j.cub.2016.01.007

Kergoat, L., Piro, B., Berggren, M., Horowitz, G., & Pham, M. C. (2012). Advances in organic transistor-based biosensors: From organic electrochemical transistors to electrolyte-gated organic field-effect transistors. *Analytical and Bioanalytical Chemistry*, *402*(5), 1813–26. https://doi.org/10.1007/s00216-011-5363-y

Kianianmomeni, A., & Hallmann, A. (2014). Algal photoreceptors: In vivo functions and potential applications. *Planta*, *239*, 1–26. https://doi.org/10.1007/s00425-013-1962-5

Kim, R. (2021). Virus as quasi-living bio-material for interaction design: Practical, ethical, and philosophical implications. In *Extended abstracts of the 2021 CHI conference on human factors in computing systems (CHI EA '21)* (pp. 1–7). Association for Computing Machinery. https://doi.org/10.1145/3411763.3451770

Kim, R., Pataranutaporn, P., Forman, J., Lee, S., Riedel-Kruse, I., Alistar, M., Lazaro Vasquez, E., Vega, K., van Dierendonck, R., Gome, G., Zuckerman, O., Vujic, A., Kong, D., Maes, P., Ishii, H., Sra, M., & Poslad, S. (2021). Microbe-HCI: Introduction and directions for growth. In *Extended abstracts of the 2021 CHI conference on human factors in computing systems (CHI EA '21)* (Article 155, pp. 1–4). Association for Computing Machinery. https://doi.org/10.1145/3411763.3450408

Kim, R., Risseeuw, C., Groutars, E., & Karana, E. (2023). Surfacing livingness in microbial displays. In *Proceedings of the 2023 CHI conference on human factors in computing systems (CHI '23)* (Article 156, pp. 1–21). Association for Computing Machinery. https://doi.org/10.1145/3544548.3581417

Kim, R., Siobhan, T., van Dierendonck, R., Bryan-Kinns, N., & Poslad, S. (2020). Working with nature's lag: Initial design lessons for slow biotic games. In *Proceedings of the 15th international conference on the Foundations of Digital Games (FDG '20)* (Article 29, pp. 1–4). Association for Computing Machinery. https://doi.org/10.1145/3402942.3409790

Kim, R., Siobhan, T., van Dierendonck, R., & Poslad, S. (2018). A new mould rush: Designing for a slow bio-digital game driven by living micro-organisms. In *Proceedings of the 13th international conference on the Foundations of Digital Games (FDG '18)* (Article 10, pp. 1–9). Association for Computing Machinery. https://doi.org/10.1145/3235765.3235798

Kim, R., van Dierendonck, R., & Poslad, S. (2019). Moldy ghosts and yeasty invasions: Glitches in hybrid bio-digital games. In *Extended abstracts of the 2019 CHI* conference *on* human factors *in* computing systems *(CHI EA '19)* (Paper LBW1619, pp. 1–6). Association for Computing Machinery. https://doi.org/10.1145/3290607.3312895

Kuznetsov, S., Taylor, A. S., Regan, T., Villar, N., & Paulos, E. (2012). At the seams: DIYbio and opportunities for HCI. In *Proceedings of the Designing Interactive Systems conference (DIS '12)* (pp. 258–267). Association for Computing Machinery. https://doi.org/10.1145/2317956.2317997

Lakatos, D., & Ishii, H. (2012). Towards radical atoms – Form-giving to transformable materials. In *Proceedings of IEEE 3rd international conference on Cognitive Infocommunications (CogInfoCom)*, Kosice, Slovakia (pp. 37–40). https://doi.org/10.1109/CogInfoCom.2012.6422023

Lee, S., Congdon, A., Parker, G., & Borst, C. (2020). *Understanding "bio" materials innovations: A primer for the fashion industry.* Fashion for Goods. Retrieved March, 2025, from https://fashionforgood.com/our_news/understanding-bio-material-innovations-a-primer-for-the-fashion-industry/

Lenau, T. A., Metzea, A.-L., & Hesselberg, T. (2018). Paradigms for biologically inspired design. In A. Lakhtakia (Ed.), *Proceedings of SPIE, international society for optical engineering* (Vol. 10593). SPIE – International Society for Optical Engineering. https://doi.org/10.1117/12.2296560

López Forniés, I., & Berges Muro, L. (2012). *International Journal of Design & Nature and Ecodynamics*, 7(1), 27–48. Retrieved March, 2025, from https://www.witpress.com/Secure/ejournals/papers/D&NE070103f.pdf

Lovley, D. R., Ueki, T., Zhang, T., Malvankar, N. S., Shrestha, P. M., Flanagan, K. A., Aklujkar, M., Butler, J. E., Giloteaux, L., Rotaru, A. E., Holmes, D. E., Franks, A. E., Orellana, R., Risso, C., & Nevin, K. P. (2011). Geobacter: The microbe electric's physiology, ecology, and practical applications. *Advances in Microbial Physiology*, *59*, 1–100. https://doi.org/10.1016/B978-0-12-387661-4.00004-5

Lovley, D. R., & Walker, D. J. F. (2019). Geobacter protein nanowires. *Frontiers in Microbiology*, *10*, 2078. https://doi.org/10.3389/fmicb.2019.02078

Mali, S. B. (2024). Bioluminescence in cancer research - Applications and challenges. *Oral Oncology Reports*, *9*, 100127. https://doi.org/10.1016/j.oor.2023.100127

Malmodin, J., & Lundén, D. (2018). The energy and carbon footprint of the global ICT and E&M sectors 2010–2015. *Sustainability*, *10*, 3027. https://doi.org/10.3390/su10093027

Mao, C., Liu, A., & Cao, B. (2009). Virus-based chemical and biological sensing. *Angewandte Chemie International Edition*, *48*(37), 6790–6810. https://doi.org/10.1002/anie.200900231

Marenko, B. (2018). The un-designability of the virtual. In G. Coombs, A. McNamara, & G. Sade (Eds.), *Undesign, critical practices at the intersection of art and design* (Chapter 3). Routledge.

Martins, T. R., Tanabe, E. H., & Bertuol, D. A. (2020). Innovative method for the recycling of end-of-life LED bulbs by mechanical processing. *Resources, Conservation and Recycling*, *161*, 104875. https://doi.org/10.1016/j.resconrec.2020.104875

McCormick, A. J., Bombelli, P., Scott, A. M., Philips, A. J., Smith, A. G., Fisher, A. C., & Howe, C. J. (2011). Photosynthetic biofilms in pure culture harness solar energy in a mediatorless Bio-Photovoltaic Cell (BPV) system. *Energy & Environmental Materials*, *4*, 4699–4709. https://doi.org/10.1039/C1EE01965A

Mimee, M., Nadeau, P., Hayward, A., Carim, S., Flanagan, S., Jerger, L., Collins, J., McDonnell, S., Swartwout, R., Citorik, R. J., Bulović, V., Langer, R., Traverso, G., & Lu, T. K. (2018). An ingestible bacterial-electronic system to monitor gastrointestinal health. *Science*, *360*(6391), 915–918. https://doi.org/10.1126/science.aas9315

Miranda, E. R., & Braund, E. (2017). Experiments in musical biocomputing: Towards new kinds of processors for audio and music. In A. Adamatzky (Ed.), *Advances in unconventional computing. Emergence, complexity and computation* (Vol. 23). Springer. https://doi.org/10.1007/978-3-319-33921-4_29

Myers, W. (2012). *Bio design: Nature, science, creativity*. Museum of Modern Art.

Nguyen, P. Q., Courchesne, N. D., Duraj-Thatte, A., Praveschotinunt, P., & Joshi, N. S. (2018). Engineered living materials: Prospects and challenges for using biological systems to direct the assembly of smart materials. *Advanced Materials*, *30*(19), e170484. https://doi.org/10.1002/adma.201704847

Niinimäki, K., Peters, G., Dahlbo, H., Perry, H., Rissanen, T., & Gwilt, A. (2020). The environmental price of fast fashion. *Nature Reviews Earth*

& *Environment*, *1*(4), 189–200. https://doi.org/10.1038/s43017-020-0039-9

Oba, Y., Konishi, K., Yano, D., Shibata, H., Kato, D., & Shirai, T. (2020). Resurrecting the ancient glow of the fireflies. *Science Advances*, *6*, eabc5705. https://doi.org/10.1126/sciadv.abc5705

Ofer, N., Bell, F., & Alistar, M. (2021). Designing direct interactions with bioluminescent algae. In *Proceedings of the 2021 ACM Designing Interactive Systems conference (DIS '21)* (pp. 1230–1241). Association for Computing Machinery. https://doi.org/10.1145/3461778.3462090

Ooms, D., Barati, B., Bruns, M., & van Dongen, T. (2022). From concern to care: A transformative reflection on designing-with the living. In *Nordic Human-Computer Interaction Conference (NordiCHI '22)* (Article 96, pp. 1–15). Association for Computing Machinery. https://doi.org/10.1145/3546155.3547285

Paley, M. A., & Prescher, J. A. (2014). Bioluminescence: A versatile technique for imaging cellular and molecular features. *MedChemComm*, *5*(3), 255–267. https://doi.org/10.1039/C3MD00288H

Parkes, A., & Dickie, C. (2013). A biological imperative for interaction design. In *CHI '13 extended abstracts on human factors in computing systems conference (CHI EA '13)* (pp. 2209–2218). Association for Computing Machinery. https://doi.org/10.1145/2468356.2468742

Pataranutaporn, P., Ingalls, T., & Finn, E. (2018). Biological HCI: Towards integrative interfaces between people, computer, and biological materials. In *CHI'18 extended abstracts on human factors in computing systems conference (CHI EA '18)* (Paper LBW579, pp. 1–6). Association for Computing Machinery. https://doi.org/10.1145/3170427.3188662

Pataranutaporn, P., Vujic, A., Kong, D. S., Maes, P., & Sra, M. (2020). Living bits: Opportunities and challenges for integrating living microorganisms in human-computer interaction. In *Proceedings of the Augmented Humans International conference (AHs '20)* (Article 30, pp. 1–12). Association for Computing Machinery. https://doi.org/10.1145/3384657.3384783

Pawar, N., Gireesh-Babu, P., Sabnis, S., Rasal, K., Murthy, R., Zaidi, S. G., Sivasubbu, S., & Chaudhari, A. (2016). Development of a fluorescent transgenic zebrafish biosensor for sensing aquatic heavy metal pollution. *Transgenic Research*, *25*(5), 617–627. https://doi.org/10.1007/s11248-016-9959-z

Perricone, V., Santulli, C., Rendina, F., & Langella, C. (2021). Organismal design and biomimetics: A problem of scale. *Biomimetics*, *6*(4), 56. https://doi.org/10.3390/biomimetics6040056

Peters, T. (2011), Nature as measure: The biomimicry, guild. *Architectural Design*, *81*, 44–47. https://doi.org/10.1002/ad.1318

Pollini, B. (2021). Sustainable design, biomimicry and biomaterials: Exploring interactivity, connectivity and smartness in nature. In

V. Rognoli & V. Ferraro (Eds.), *ICS materials interactive, connected, and smart materials* (pp. 60–73). Franco Angeli.

Poupyrev, I., Schoessler, P., Loh, J., & Sato, M. (2012). Botanicus interacticus: Interactive plants technology. In *Proceedings of ACM SIGGRAPH 2012 emerging technologies (SIGGRAPH '12)* (Article 4, p. 1). Association for Computing Machinery. https://doi.org/10.1145/2343456.2343460

Prindle, A., Samayoa, P., Razinkov, I., Danino, T., Tsimring, L., & Hasty, J. (2012). A sensing array of radically coupled genetic 'biopixels'. *Nature*, *481*, 39–44. https://doi.org/10.1038/nature10722

Provin, A., de Aguiar Dutra A. R., Machado M., & Vieira Cubas A. (2021). New materials for clothing: Rethinking possibilities through a sustainability approach–a review. *Journal of Cleaner Production*, *282*, 124444. https://doi.org/10.1016/j.jclepro.2020.124444

Quijano, L., Speight, R., & Payne, A. (2021). Future fashion, biotechnology and the living world: Microbial cell factories and forming new 'oddkins'. *Continuum*, *35*(6), 897–913. https://doi.org/10.1080/10304312.2021.1993574

Rathinamoorthy, R., Suvitha, L., & Raja Balasaraswathi, S. (2023). Sustainable technologies and materials for future fashion. In S. S. Muthu (Ed.), *Novel sustainable process alternatives for the textiles and fashion industry. Sustainable textiles: Production, processing, manufacturing & chemistry*. Springer. https://doi.org/10.1007/978-3-031-35451-9_5

Rebatchouk, D., Daraselia, N., & Narita, J. O. (1996). NOMAD: a versatile strategy for in vitro DNA manipulation applied to promoter analysis and vector design. *Proc Natl Acad Sci USA*, *93*, 10891–6. https://doi.org/10.1073/pnas.93.20.10891

Risseeuw, C., Martinez Castro, J. F., Barla, P., & Karana, E. (2023). FlavoMetrics: Towards a digital tool to understand and tune living aesthetics of flavobacteria. In *DIS '23: Proceedings of the 2023 ACM designing interactive systems conference* (pp. 2079–2092). Association for Computing Machinery (ACM). https://doi.org/10.1145/3563657.3596085

Rognoli, V., Petreca, B., Pollini, B., & Saito, C. (2022). Materials biography as a tool for designers' exploration of bio-based and bio-fabricated materials for the sustainable fashion industry. *Sustainability: Science, Practice and Policy*, *18*(1), 749–772. https://doi.org/10.1080/15487733.2022.2124740

Saha, D., Mehta, D., Altan, E., Chandak, R., Traner, M., Lo, R., Gupta, P., Singamaneni, S., Chakrabartty, S., & Raman, B. (2020). Explosive sensing with insect-based biorobots. *Biosensors and Bioelectronics: X*, *6*, 100050. https://doi.org/10.1016/j.biosx.2020.100050

Sareen, H., Zheng, J., & Maes, P. (2019). Cyborg botany: Augmented plants as sensors, displays and actuators. In *Extended abstracts of the 2019 CHI* conference *on* human factors *in* computing systems *(CHI EA '19)*

(Paper VS13, pp. 1–2). Association for Computing Machinery. https://doi.org/10.1145/3290607.3311778
Sawa, M., Fantuzzi, A., Bombelli, P., Howe, C. J., Hellgardt, K., & Nixon, P. J. (2017). Electricity generation from digitally printed cyanobacteria. *Nature Communications*, *8*, 1327. https://doi.org/10.1038/s41467-017-01084-4
Seidu, R. K., Eghan, B., & Acquaye, R. (2023). A review of circular fashion and bio-based materials in the fashion industry. *Circular Economy and Sustainability*. Advance online publication. https://doi.org/10.1007/s43615-023-00303-z
Seznec, Y., & Hickey, P. (2013). The Secret Sounds of Spores [Installation]. Retrieved from https://www.yannseznec.com/docs/projects/spores/
Shamoon, A., Haleem, A., Bahl, S., Javaid, M., Garg, S. B., Sharma, R. C., & Garg, J. (2022). Environmental impact of energy production and extraction of materials - A review. *Materials Today: Proceedings*, *57*(Part 2), 936–941. https://doi.org/10.1016/j.matpr.2022.03.159
Shang, L., Shao, C., Chi, J., & Zhao, Y. (2021). Living materials for life healthcare. *Accounts of Materials Research*, *2*(1), 59–70. https://doi.org/10.1021/accountsmr.0c00084
Sharifian, S., Homaei, A., Hemmati, R., Luwor, B., & Khajeh, K. (2018). The emerging use of bioluminescence in medical research. *Biomedicine & Pharmacotherapy*, *101*, 74–86. https://doi.org/10.1016/j.biopha.2018.02.065.
Shlosberg, Y., Schuster, G., & Adir, N. (2022). Self-enclosed bio-photoelectrochemical cell in succulent plants. *ACS Applied Materials & Interfaces*, *14*(48), 53761–53766. https://doi.org/10.1021/acsami.2c15123
Smolke, C. D. (2009). Building outside of the box: iGEM and the BioBricks Foundation. *Nature Biotechnology*, *27*(12), 1099–1102. https://doi.org/10.1038/nbt1209-1099
Srivastava, A., & Katiyar, K. (2021). The ecology of bioluminescence. In H. Suzuki & K. Ogoh (Eds.), *Bioluminescence - Technology and biology*. IntechOpen. https://doi.org/10.5772/intechopen.96636
Steenmeijer, M. A., Rodrigues, J. F. D., Zijp, M. C., & Loop, S. L. W. (2022). The environmental impact of the Dutch health-care sector beyond climate change: An input–output analysis. *The Lancet Planetary Health*, *6*(12), e949–e957. https://doi.org/10.1016/S2542-5196(22)00244-3
Stępień, J. (2021). Performing with/for the algae world in the Anthropocene. Burton Nitta's Algaculture projects. *International Journal of Performance Arts and Digital Media*, *17*(3), 303–315. https://doi.org/10.1080/14794713.2020.1816701
Travis, J. (2011). Don't call it viral marketing: The story behind contagion's microbial billboard. *Science*. Retrieved March, 2025, from https://www.science.org/content/article/dont-call-it-viral-marketing-story-behind-contagions-microbial-billboard

Van Dortmont, T. (2019). *Growing light* [Master's thesis, Delft, The Netherlands]. Retrieved March, 2025, from https://repository.tudelft.nl/islandora/object/uuid%3A14fdd5ac-e077-403c-992a-d40a526482cf

Von Thaden, J., Badillo-Montaño, R., Lira-Noriega, A., García-Ramírez, A., Benítez, G., Equihua, M., Looker, N., & Pérez-Maqueo, O. (2021). Contributions of green spaces and isolated trees to landscape connectivity in an urban landscape. *Urban Forestry & Urban Greening*, *64*, 127277. https://doi.org/10.1016/j.ufug.2021.127277

Wallace Luz. (2024). Lichenized biosystems. Retrieved February, 2025, from https://www.pratt.edu/work/lichenized-biosystems/

Wilson, T. (2013). *Bioluminescence: Living lights, lights for living*. Harvard University Press. https://doi.org/10.4159/harvard.9780674068025

Yao, L., Ou, J., Cheng, C., Steiner, H., Wang, W., Wang, G., & Ishii, H. (2015). bioLogic: Natto cells as nanoactuators for shape changing interfaces. In *Proceedings of the 33rd annual ACM conference on human factors in computing systems (CHI '15)* (pp. 1–10). Association for Computing Machinery. https://doi.org/10.1145/2702123.2702611

Yuk, H., Lu, B., & Zhao, X. (2019). Hydrogel bioelectronics. *Chemical Society Reviews*, *48*, 1642–1667. https://doi.org/10.1039/C8CS00595H

Zhang, Y., Hsu, L. H. H., & Jiang, X. (2020). Living electronics. *Nano Research*, *13*, 1205–1213. https://doi.org/10.1007/s12274-019-2570-x

Zhou, J., Barati, B., Giaccardi, E., & Karana, E. (2022). Habitabilities of living artefacts: A taxonomy of digital tools for biodesign. *International Journal of Design*, *16*(2), 57.

Zhou, J., Kim, R., Doubrovski, Z., Martins, J., Giaccardi, E., & Karana, E. (2023). Cyano-chromic interface: Aligning human-microbe temporalities towards noticing and attending to living artefacts. In *Proceedings of the 2023 ACM Designing Interactive Systems conference (DIS '23)* (pp. 820–838). Association for Computing Machinery. https://doi.org/10.1145/3563657.3596132

Zhou, X., Zhou, D., Wang, Q., & Su, B. (2019). How information and communication technology drives carbon emissions: A sector-level analysis for China. *Energy Economics*, *81*, 380–392. https://doi.org/10.1016/j.eneco.2019.04.014

CHAPTER 4

TOWARDS DILS

MAPPING INTERACTIONS WITH BIO-UX

DOI: 10.4324/9781003366799-5

FOUNDATIONAL CONCEPTS AND PROMISING PERSPECTIVES

Biodesign, an emerging field at the intersection of design, biology and technology, marks a pivotal moment in the evolving relationship between humans and nature. It introduces a fundamental paradigm shift by moving beyond anthropocentric frameworks to a holistic perspective that recognizes the interconnectedness of all species – including non-human ones (Rosén et al., 2022; Willis, 2006; Braidotti, 2013). In this view, design no longer aims solely to meet the needs of human users but also considers the well-being of all organisms within the broader ecosystem.

This radical redefinition of the design process has significant implications across multiple domains – from the dynamics of user-device interaction to how designers perceive and engage with systems, materials and tools. The development of biodesign seeks to transcend conventional human-centered and technology-focused paradigms in favor of symbiotic design practices. In this context, technological artifacts are no longer seen as isolated products but as elements of inclusive, regenerative systems that respect and sustain the diversity of life (Gough et al., 2020).

Such a paradigm encourages designers to move beyond biologically inspired aesthetics and metaphors to practices that incorporate living organisms as active components. At the convergence of biology and technology, new design pathways are emerging that foster ecological awareness, enable multispecies agency and support biodiversity. This approach marks a significant shift in how sustainable, life-centered innovations for the future are conceived and developed.

A crucial theoretical advance in this area is the concept of *Living Bits* (Pataranutaporn et al., 2018), which identifies five distinct categories that describe the role of microorganisms as interfaces in human–computer interaction (HCI). This framework strengthens the integration of biological elements into interactive systems, opening up new possibilities for interface design.

Equally important is the concept of *livingness* as a material quality (Karana et al., 2020), which introduces biological, ecological and experiential dimensions into product design. It supports methodologies that incorporate notions of *living aesthetics*, *mutualistic care* and *habitability*. The last is particularly important, as it emphasizes designing for the viability and well-being of humans and non-humans throughout the lifecycle of a living artifact.

Supporting this perspective, the *Habitability Taxonomy* proposed by Zhou et al. (2022) provides a digital toolkit for understanding, integrating and maintaining designed habitats. It highlights the designer's role across the biodesign continuum, from configuring biological habitats during the design phase to facilitating their social and environmental maintenance over time.

Despite the growing interest in biodesign and the proliferation of experimental projects promoting regenerative design and multispecies coexistence, the field still lacks empirical research into the practical dynamics between products, people and living organisms. Much of the current work remains theoretical or speculative.

However, recent advances in life sciences, biotechnology and synthetic biology are creating promising new scenarios. They enable the creation of artifacts beyond traditional notions of human–computer or human–object interaction (Bell et al., 2024). These pioneering artifacts – involving living entities such as algae, bacteria or fungi – require a transformation in UX methodologies and interaction design tools, capable of accommodating the complexity, unpredictability and care inherent in designing with and for living systems.

In this context, critical reflection on the methodologies, languages and tools of contemporary interaction design is essential to responsibly navigating this new frontier.

This shift toward the regenerative and symbiotic design of DILS requires a re-examination of the principles and tools traditionally used in user experience and interaction design. While these fields have long prioritized usability, efficiency and emotional engagement, integrating living systems introduces new levels of complexity, responsiveness and care. To fully engage with this emerging paradigm, it is essential to understand the fundamentals of UX and interaction design and the mapping tools that guide designers in analyzing and shaping user interactions. These frameworks are the foundation upon which biointegrated, multispecies interactions can be developed.

User Experience Design (UX)

User experience design (UX design) is a user-centered approach to product development that seeks to align the needs and expectations of users with the goals of designers and developers. As an interdisciplinary field, it focuses on creating products, interfaces, services and systems that enhance and prioritize the overall user experience.[1] In both physical and digital contexts, UX design refers to the process of creating artifacts that provide meaningful and relevant experiences. This includes the product itself and the broader journey of discovery, acquisition, interaction and integration, which touches on aspects such as branding, usability, design and functionality. UX design considers each stage of this journey to ensure intuitive, efficient and satisfying interactions across platforms, including websites, mobile applications, connected objects and technological services (Norman & Nielsen, 2002; Moggridge, 2007).

Embracing usability, interaction design, information architecture and emotional engagement, UX explores how users perceive and interact with products over time (Garrett, 2011; Norman, 2013). This holistic perspective ensures consistency and coherence, establishing UX as a pillar of contemporary product development.

At its core, UX design requires a deep understanding of users – their capabilities, goals, usage contexts and potential limitations (Bødker & Klokmose, 2019). The design process begins with empathetic, user-centered research to capture experiential nuances, followed by an iterative cycle of ideation, prototyping, testing and refinement, with each phase informed by direct user feedback.

The main goal of digital UX design is to ensure usability and ease of navigation, often through intuitive interfaces, clear pathways, and reduced cognitive load and frustration. UX designers structure content and interactions logically to help users achieve their objectives effectively.

Importantly, UX also balances aesthetics with functionality. It prioritizes user needs over technical or stylistic choices, aiming to deliver experiences that are not only usable but also emotionally resonant. Prototypes are rigorously tested and refined based on feedback, reinforcing UX's role in fostering long-term user satisfaction (Hassenzahl, 2010).

Interaction Design

Interaction design is a specialized area within UX design that focuses on shaping how users engage with interactive systems. It defines how a product responds to human user input – such as button clicks, link navigation, animation feedback or interaction flows within applications and websites. Its primary goal is to create systems that enable users to achieve their goals intuitively, efficiently and satisfactorily.

While UX design considers the entire user journey and emotional experience, interaction design focuses on the quality of the moments of interaction between the user and the product. The two are interdependent: excellent UX cannot exist without well-developed interaction design, and interaction design depends on the strategic goals set by UX.

Interaction designers use tools and mapping techniques to visualize, analyze and communicate user interactions to support this. These include user flow diagrams, wireframes and touchpoint maps, which help identify key interaction moments, sources of frustration, and opportunities for improvement (King & Chang, 2016; Saffer, 2009).

The following are some of the main mapping tools used in interaction design.

User Journey Maps[2]

These visual tools illustrate a human user's step-by-step process to achieve a specific goal when interacting with a product or service. They capture the user's actions, thoughts, emotions and motivations throughout the experience, allowing designers to identify key touchpoints, pain points and moments of satisfaction.

By providing a holistic view of user behavior, expectations and potential challenges, journey maps help design teams gain deeper insight into the user experience. This comprehensive understanding supports the identification of critical areas for improvement and guides the creation of more intuitive, engaging and user-centric solutions (Endmann & Keßner, 2016).

Experience Maps[3]

Like user journey maps, experience maps visualize the user experience but take a broader and more strategic view. While journey maps focus on specific interactions with a product or service, experience maps consider the entire user journey – including overarching goals, emotional states and contextual factors that influence behavior.

Their primary purpose is to capture the full spectrum of user engagement, from initial awareness and discovery through interaction and post-use reflection. By placing the product or service in the broader context of the user's life, experience maps reveal deeper insights into perceptions, needs and motivations beyond isolated touchpoints.

This comprehensive view enables design teams to uncover richer opportunities for innovation and develop empathetic and user-centric solutions. By understanding how users engage and why they behave in specific ways, experience maps support the creation of meaningful, contextualized experiences (Szabo, 2017).

Empathy Maps[4]

Empathy maps are powerful visual tools that help design teams gain a deeper, more human-centered understanding of users. They are particularly valuable for fostering empathy and ensuring the design process focuses on real user needs and lived experiences.

These maps typically divide the user experience into four key dimensions: what the user says, does, thinks and feels. This structure encourages teams to consider observable behaviors and the underlying emotions, motivations and perceptions that may not be immediately apparent.

By aligning the design team's perspective with that of the user, empathy maps enable more informed, sensitive and user-centered design decisions. They help ensure that the final products or services resonate with users on both a practical and emotional level (Ferreira et al., 2015).

Storyboarding[5]

Storyboarding is a visual storytelling technique widely used in design and filmmaking to outline a sequence of events or interactions. In the context of UX and interaction design (ID), it allows designers to represent the user journey in a narrative format, capturing interactions, experiences and emotional shifts over time.

Storyboards typically consist of illustrated or sketched frames, each representing a specific moment or action within the user experience. Arranged chronologically, these frames convey the interaction flow and place user behavior in a realistic context.

This method is particularly valuable for communicating the context of use, exploring design ideas and highlighting emotional touchpoints. By translating user experiences into visual narratives, storyboarding helps design teams better understand needs, identify pain points and uncover opportunities for improvement. It also facilitates interdisciplinary collaboration by expressing aspects of the user experience that are difficult to capture in text or data alone (Truong et al., 2006).

BEYOND THE HUMAN: BIO-UX AS A METHODOLOGY FOR LIVING INTERACTION DESIGN

Traditionally, interaction design has centered on the human user. The design process has focused on understanding and responding to human needs, goals and constraints from research to testing. However, this perspective changes radically when another living being becomes part of the interaction. In the Designed Interactive Living Systems (DILS) context, humans are no longer the only actors; living organisms co-shape function and meaning. The approach becomes bifocal – balancing human-centerd priorities with the needs, behaviors, and well-being of non-human organisms involved in the design process.

This transformative shift requires a rethinking of user experience design through a bio-integrated lens, requiring a fundamentally new way of thinking that benefits not only human users but also the organisms involved in the interaction and the ecosystems in which these interactions occur. To support this evolution, we propose a new concept within the field of user-centered design: bio-UX design, a methodology that extends traditional UX to include a multispecies perspective. Rather than designing for the user alone, bio-UX proposes designing with and through living systems – acknowledging their agency, limitations and ecological role. It seeks to draw parallels between what defines and guides human users and the characteristics, behaviors and needs of non-human life forms. As Albergati (2020) notes, this orientation is inspired by the collaborative intelligence of nature itself.

Within the broader framework of living interactions design, bio-UX offers practical tools for navigating the complexities of designing for regenerative ecologies. As interactive systems begin to incorporate living organisms, it is becoming increasingly clear that traditional UX methodologies must be rethought to accommodate biological collaborators' characteristics, capabilities and limitations. A series of original maps have been developed to extend established UX frameworks in response to this need. These tools support the design of more ethical, adaptive and symbiotic interactions between humans and non-human living entities, positioning bio-UX as a foundational methodology at the core of DILS.

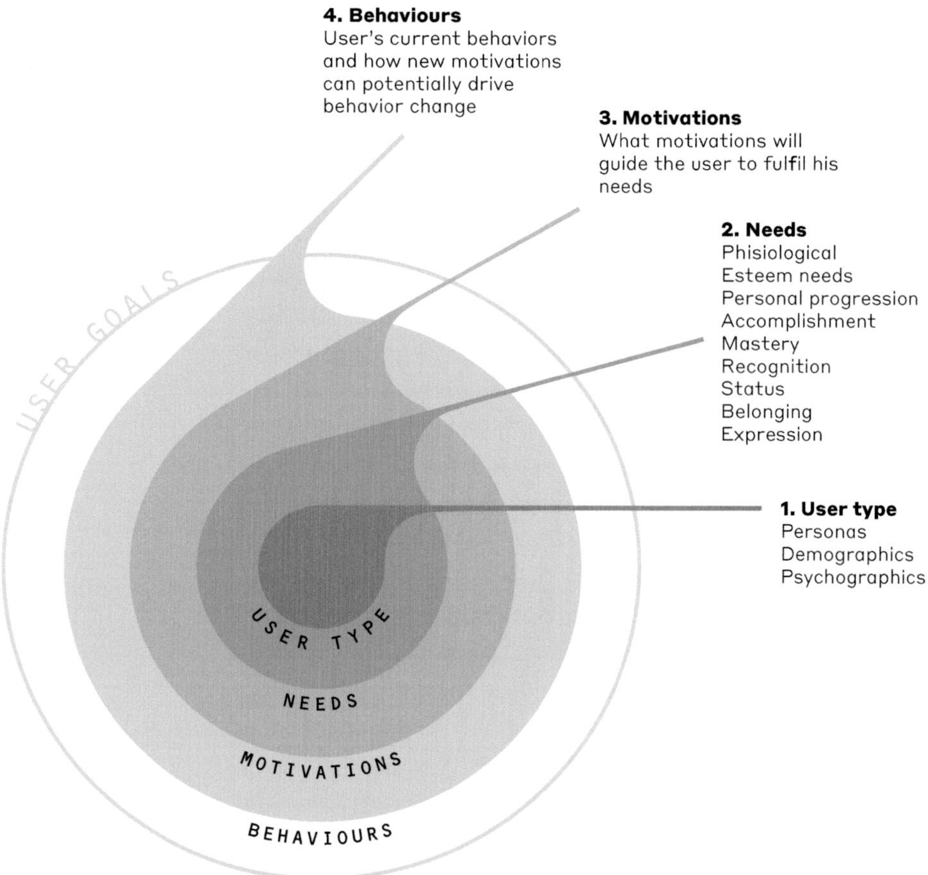

Figure 4.1 User Goals Map (2024). A four-level framework for understanding the user perspective, illustrating the relationships between user types, needs, motivations, and behaviors.
Credits: Elena Albergati, adapted from Corey Stern's User Goals Map (2014).

User Goals Map

In the early stages of a user experience design project, it is essential to conduct a comprehensive analysis of the end users who will interact with the product or service. Designers often rely on various mapping tools selected and adapted according to the specific goals and context of the project (Figure 4.1).

The User Goals Map presented here is inspired by a framework developed in 2014 by UX designer Corey Stern, who combined the strengths of various user analysis diagrams into a unified and versatile model. This map is structured around four key levels: *User Type, Needs, Motivations and Behaviors.*

1. *User types*: Understanding the different roles that users may take on is critical to effective design. A standard method is to develop user personas that encapsulate these roles, detailing responsibilities, skill levels, demographics (e.g. gender, age, language, location) and psychographics (e.g. personality, values, attitudes, interests, lifestyle). This step also considers when, where and how users engage with the product.
2. *Needs*: Once user types have been identified, it is crucial to identify and define the needs and wants that users need to meet to achieve their goals. These needs can range from the basic, such as access to documentation, to the more complex, including physiological, emotional or esteem-related dimensions. Examples include personal development, achievement, recognition, belonging, self-expression and a sense of purpose.
3. *Motivations*: Having identified user needs, the next step is to examine the underlying motivations that drive users to satisfy them. These can include a variety of intrinsic and extrinsic incentives that can be activated through design cues, feedback mechanisms or behavioral triggers.
4. *Behaviors*: Understanding current user behavior – and how it might evolve in response to new motivations – is essential. Strong motivation often leads to behavioral change, increasing the likelihood of habit formation and long-term engagement with the product or brand.

Once analyzed, these four dimensions provide designers with actionable, measurable insights to create more meaningful, effective and user-centric experiences that enhance user well-being.

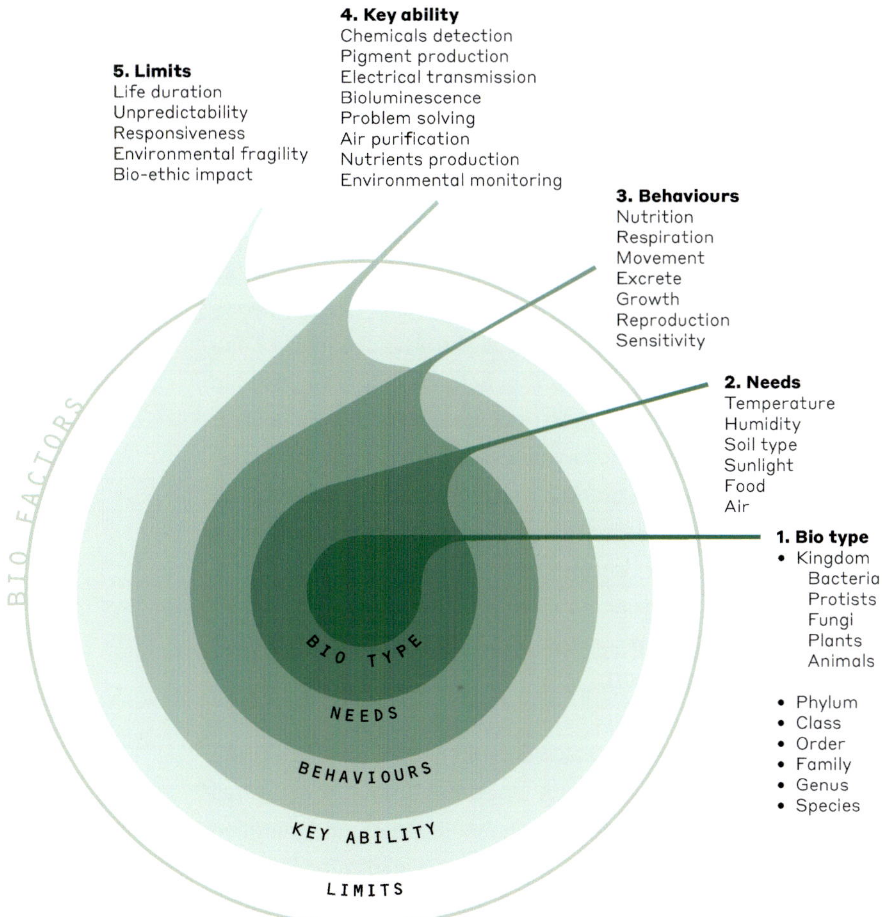

Figure 4.2 Bio Factors Map (2024). A five-level framework designed to analyze the biological and behavioral characteristics of living organisms integrated into design projects. The map highlights five key dimensions: biotype, needs, behavior, key capabilities, and limits.
Credits: Elena Albergati.

Bio Factors Map

The successful integration of living organisms into designed products requires a comprehensive understanding of their defining characteristics. This knowledge is essential to identify the organisms' capabilities and limitations, set realistic expectations and determine the most appropriate collaboration methods (Figure 4.2). A tailored tool – the Bio Factors Map – has been developed to support this process. This model has five key levels: *Biotype, Needs, Behavior, Communication and Limits.*

1. Biotype

   The first level classifies the organism within the taxonomic hierarchy: *Kingdom, Phylum, Class, Order, Family, Genus and Species.* The first phase is the identification of the kingdom to which the organism belongs, with five possible kingdoms: *Monera* (which includes microscopic single-celled organisms, prokaryotes, such as various bacteria), *Protista* (a diverse group of single-celled organisms with nutritional characteristics like plants (algae) or fungi and animals (amoebae, foraminifera, radiolarians)), *Fungi, Plants* or *Animals.* The specific species of the organism is then determined based on its distinguishing characteristics (Margulis et al., 2009).

2. Needs

   All living organisms have basic survival needs, although these vary significantly between species. They generally fall into favorable external conditions (habitat) and nutrition (food). Habitat includes temperature, humidity, soil quality, sunlight and air quality. Microorganisms utilize a diverse array of substances as "food" or nutritional sources, tapping into various organic and inorganic materials to fulfil their metabolic requirements (Kawade, 2009). Accurate assessment of these needs ensures the organism's viability within the designed product.

3. Behaviors

   All living organisms exhibit seven fundamental life processes: nutrition, respiration, movement, excretion, growth, reproduction and sensitivity (Campbell & Reece, 2005):

   - *Nutrition*: Acquisition of nutrients from their environment (e.g. proteins, carbohydrates, fats) for growth and energy.

- *Respiration*: Conversion of nutrients into usable energy via aerobic or anaerobic pathways. Respiration is essential for various cellular processes, ensuring cells have the energy to maintain, grow and reproduce. Two primary types of respiration occur: aerobic respiration (an energy-generating process in cells using oxygen to convert nutrients into adenosine triphosphate (ATP), carbon dioxide and water) and anaerobic respiration (an energy-generating process in cells without using oxygen, instead employing alternative electron acceptors to convert nutrients into ATP and various byproducts).
- *Movement*: Varying degrees of movement depending on the organism.
- *Excretion*: Elimination of metabolic waste.
- *Growth*: Development and multiplication of cells, usually facilitated by the consumption of food for cell production.
- *Reproduction*: The ability to produce offspring.
- *Sensitivity*: Sense and respond to external stimuli such as light, temperature or chemicals.

4. Key capabilities

   Once the behavioral traits have been identified, it is necessary to determine the organism's most important and unique capabilities and functions. These may include, for example, the ability of bacteria to detect pollutants, the potential of algae for air purification, the information transfer properties of mycelium or the problem-solving ability of *Physarum polycephalum*, just to name a few. Understanding these can inform domain-specific applications.

5. Limits

   In addition to capabilities, designers must carefully assess the limits of the organism to avoid unnatural or non-viable integration.

   Key considerations for the design process include:

   - *Lifespan*: No organism lives indefinitely; continuous feeding or periodic replacement is required.
   - *Natural variability*: Living organisms are inherently unpredictable, creating challenges for standardization.

- *Responsiveness*: Consider the organism's ability to respond to certain stimuli while remaining unresponsive to others.
- *Bioethics*: Risks and ethical implications must be rigorously assessed, especially when using genetic modification. Involving microorganisms, mainly bacteria, genetic engineering or modification may confer capabilities that are not naturally present. Evaluate these modifications based on risk–benefit analysis and ethical principles. This will ensure not forcing the organism to behave unnaturally, avoiding the creation of non-functional products.

By methodically exploring these five dimensions using the Bio Factors Map, designers can make informed, ethical and effective decisions about integrating living organisms into interactive products for creating a DILS.

Bio-User Interaction Map

After analyzing the distinct characteristics of the human user and the organism, their intersection can be explored through a structured framework that defines how their interaction unfolds (see Figure 4.3). The Bio-User Interaction Map introduces six key

Figure 4.3 Bio-User Interaction Map (2024). Visualization of the intersection between User Goals and Bio Factors, illustrating the shared space where human and biological needs align.
Credits: Elena Albergati.

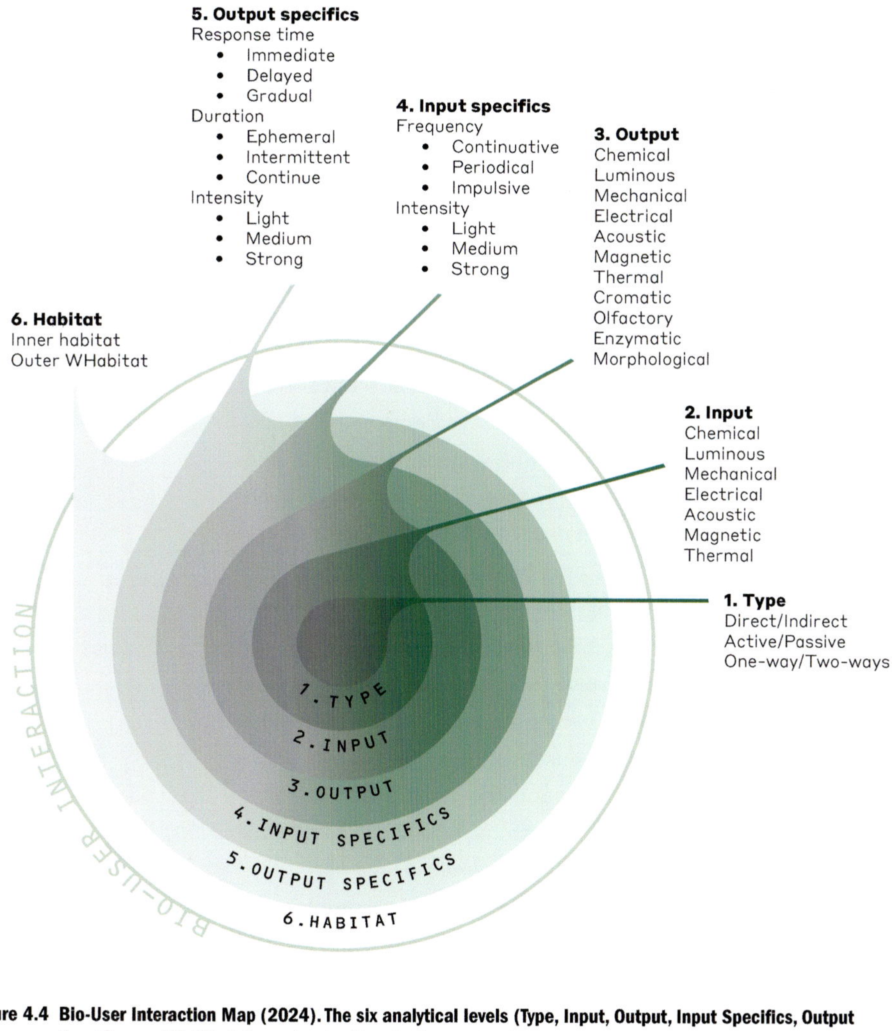

Figure 4.4 Bio-User Interaction Map (2024). The six analytical levels (Type, Input, Output, Input Specifics, Output Specifics, and Habitat) for understanding the interaction between the human user and the living organism. Credits: Elena Albergati.

dimensions: Type, Input, Output, Input Specifics, Output Specifics and Habitat (see Figure 4.4).

- *Type of interaction*: Interactions between living beings can be direct – where the human user engages with the organism without mediation – or indirect, where technological components such as sensors or software are involved. For example, interaction with bioluminescent algae is direct because no interface mediates the light emission. Conversely, in *The Secret Sound of Spores*,[6] software translates the movement of spores into sound, making the interaction indirect.

Interactions can also be active, that is, initiated by the human user who deliberately triggers the organism's response, or passive, that is, the human user is merely a spectator witnessing the organism's response to external stimuli. In *Mould Rush*,[7] the user directly influences fungal growth – an active interaction. In contrast, Julian Melchiorri's *Exhale*[8] invites passive observation of algae purifying the air through photosynthesis.

Finally, interactions are either one-way or two-way. In one-way interactions, the organism reacts to the human user but does not influence the human user in return. The interaction in the *Living Light Dress*[9] project is one-way because the user's movement generates the algae's light. Still, the algae's light does not influence the user's behavior in return.

Two-way interactions involve feedback loops, meaning that not only does the human user's input affect the organism, but the organism's feedback also affects the human user's actions. In the *E.Chromi*[10] project, the human user's health status affects the colors of the bacteria, and at the same time, based on the colors obtained, the human users can adjust their actions, such as addressing the identified problem, so the interaction is two-way.

- *Input*: Interaction requires the organism to receive input from the human user or the environment. Living organisms can respond to a wide range of stimuli – chemical, light, mechanical, acoustic, electrical and magnetic (e.g. pressure, stretch, vibration, oscillation or rotation). Unlike traditional human–computer interaction (HCI), where inputs are primarily visual, auditory or tactile, biological interaction requires a broader sensitivity to environmental cues. Understanding how organisms perceive and respond to these stimuli is essential for designing meaningful and functional interactions with DILS.

- *Output*: Output refers to the response produced by the organism in response to a given input. Depending on the species and stimulus, this response can take many forms – light, mechanical, chemical, chromatic, acoustic, thermal, olfactory, enzymatic, morphological, magnetic or electrical. Understanding how the organism communicates is critical to designing effective interactions. Unlike digital systems, which typically rely on limited outputs such as visual, acoustic or haptic signals, living organisms exhibit a broader and more diverse range of responses.

  For example, bioluminescent algae produce light (light output) in response to mechanical stimulation (e.g. movement); microalgae perform photosynthesis under light (light input), releasing oxygen (chemical output); certain bacteria change color when they detect certain substances (chemical input, chromatic output). These examples illustrate the complexity and richness of biological outputs, which must be carefully interpreted and integrated into the design of living interactive systems.
- *Input specifics*: Understanding how an organism responds to stimuli variations is essential to designing interactive living systems. Stimuli can be classified by frequency – impulsive (a single action), periodic (repeated actions) or continuous (uninterrupted stimulation) – and by intensity, which can range from mild to moderate or strong. These characteristics significantly influence the organism's ability to produce a meaningful response.

  For example, bioluminescent algae will only emit light in response to moderate or strong mechanical stimulation. If the movement is too gentle, there will be no visible output. In addition, they cannot sustain continuous stimulation; after about 2 minutes of constant agitation, they require a rest period of about 30 minutes before they can glow again. Short, impulsive movements trigger only a brief light response, while periodic, moderate inputs allow repeated activations without exhausting the organism.

Designers must assess the appropriate type and intensity of input required to elicit and maintain a desired biological response. Matching these parameters to the organism's capabilities will ensure a more reliable, ethical and effective interaction.

- *Output specifics*: Understanding the detailed characteristics of an organism's response is essential to designing meaningful interactions. Outcomes vary along several dimensions: timing (immediate, delayed, or gradual), duration (temporary, intermittent, or continuous), and intensity (ranging from mild to strong). These factors significantly influence the design potential and applicability of the interaction – especially in contexts such as real-time feedback, prolonged use or sensory engagement.

  For example, bioluminescent algae respond to mechanical stimulation with an immediate and transient light emission. The intensity of their glow depends on the strength of the stimulus – a mild input results in a weak response, while stronger agitation produces a more vivid light. Notably, some organisms can respond to a wide range of stimuli but only show a significant output in response to specific, well-calibrated conditions. Recognizing these nuances allows designers to match interaction goals with the biological capabilities of the organism involved.

- *Habitat*: Effective human user–organism interaction requires a comprehensive understanding of the organism's relationship with its environment. This involves considering two interconnected habitats: the internal habitat, which refers to the internal conditions within the product that support and maintain the organism (e.g., temperature, nutrients, enclosure); and the external habitat, which is the external context in which the product is used and interacts with its environment.

  Maintaining both environments is essential to ensure the organism's well-being and enable functional interaction. For example, in the *Living Light Dress*, the inner habitat consists of the bioreactive chamber embedded in the garment where the algae reside. At the same time, the outer habitat is defined by the user's body moving in the surrounding space, which influences the interaction dynamics (Karana et al., 2020).

By carefully considering six key dimensions – *interaction type, degree, input, output, specifications and habitat* – designers can cultivate more effective, ethical and context-sensitive interactions between users and living organisms. Unlike conventional interfaces, which often rely on static, predefined modalities, interactions within DILS require a more fluid, dynamic approach that embraces biological variability and fosters mutual care.

This expanded perspective recognizes the fundamental differences between digital and biological interactions and emphasizes the importance of designing functional, experientially and emotionally engaging systems. The interplay of direct and indirect stimuli, active and passive engagement, and variable response intensities shape the overall interaction experience, influencing both the usability and depth of the relationship.

Equally critical is the ethical and environmental responsibility inherent in designing with living systems. Properly conceived interactions support the longevity and well-being of humans and organisms, while inadequate designs risk causing harm, exploitation or system failure. The Bio-User Interaction Map is emerging as a strategic tool for navigating these complexities, enabling designers to foster meaningful, regenerative relationships and support the transition to more symbiotic design practices within the broader framework of DILS.

Symbiosis Map

Building on the Bio-User Interaction Map, which focuses on the relationship between users and organisms, the Symbiosis Map adopts a different perspective, shifting the focus towards mutual

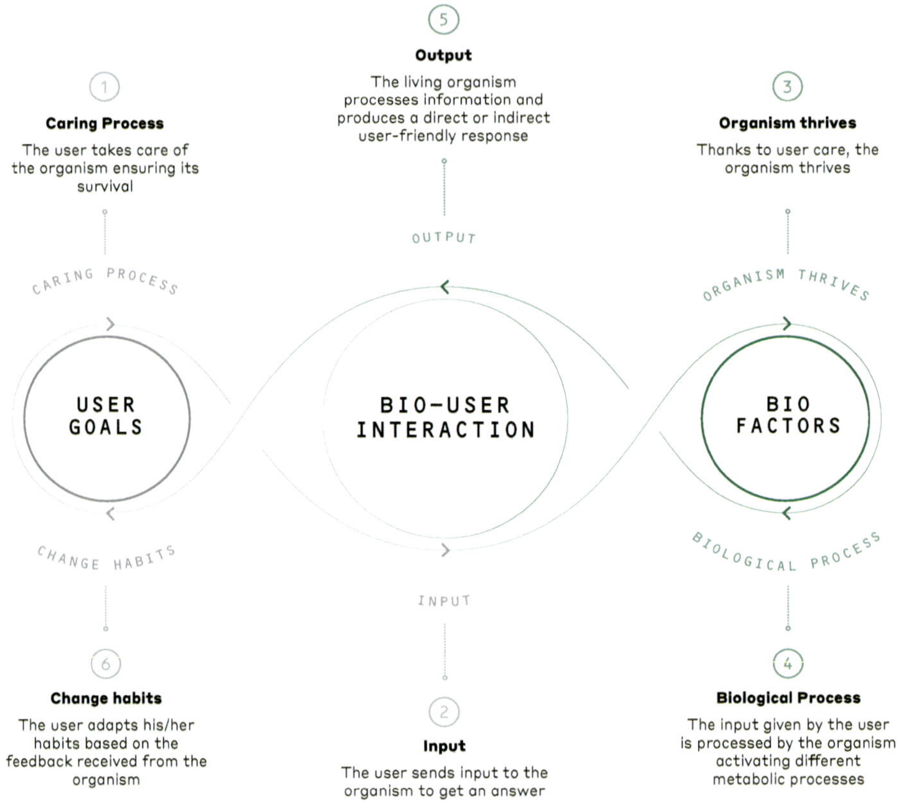

Figure 4.5 Symbiosis Map (2024). Framework illustrating how designers can balance the needs of the human user and the living organism through mutual care. The map visualizes the reciprocal flow between user goals and bio factors across six iterative steps: Care Process, Input, Biological Process, Organism Thrives, Output, and Change Habits.
Credits: Elena Albergati.

care and co-dependence and framing interaction as a continuous, reciprocal process. It provides a framework for exploring how designers can balance the needs of human users and organisms effectively (see Figure 4.5). It visually maps the intersection between user goals and biofactors through the lens of mutual care (symbiosis), placing meaningful interaction at its core. While all products that incorporate living organisms require some level of care, a truly symbiotic relationship between user and organism is not always guaranteed. In some cases, the product itself – rather than the human user – manages and maintains the living component.

This map is designed to analyze scenarios with mutual influence: the human user's actions affect the organism, and the organism's responses affect the user. By foregrounding this reciprocal dynamic, the Symbiosis Map encourages designers to foster harmonious, co-beneficial interactions that enrich both the experience and the outcome.

The structure of the map includes four key components: User Goals (left), Bio Factors (right), Bio-User Interaction (center), and the Symbiosis Flow that connects them. While User Goals and Bio Factors have been detailed previously, this section focuses on the Symbiosis Flow – an iterative sequence of six steps: *Care Process*, *Input*, *Biological Process*, *Organism Thrives*, *Output and Change Habits.*

- *Caring process*: Interaction begins with care, which is essential for the organism's survival. The human user plays a vital role by providing food, removing waste, maintaining a clean environment and providing adequate light. These combined actions allow the organism within the product to thrive and extend its life over time.
- *Input*: The human user engages the organism through specific stimuli, initiating the interaction.
- *The organism thrives*: In response to consistent care, the organism enters a thriving state – absorbing nutrients, breathing, growing, reproducing, cleansing or moving. This state enables it to perceive and process environmental stimuli actively, providing the basis for meaningful interaction.
- *Biological process*: The organism metabolizes the input it receives from the human user through various biological and metabolic functions, activating processes that lead to a specific response – its output.

- *Output*: The organism processes the input received from the human user through various biological and metabolic processes. The body interprets these stimuli and generates a response called the output.
- *Change habits*: Based on the organism's response, the human user interprets the feedback and may adjust their behavior accordingly, completing the feedback loop and reinforcing a mutually responsive relationship.

Project Flow

Once the needs of users and organisms have been analyzed and their interactions understood, the next step is to embed these findings within a broader design process. The process, called Project Flow presented here, adapts the classical design thinking framework proposed by Sarah Gibbons (2016), restructured to

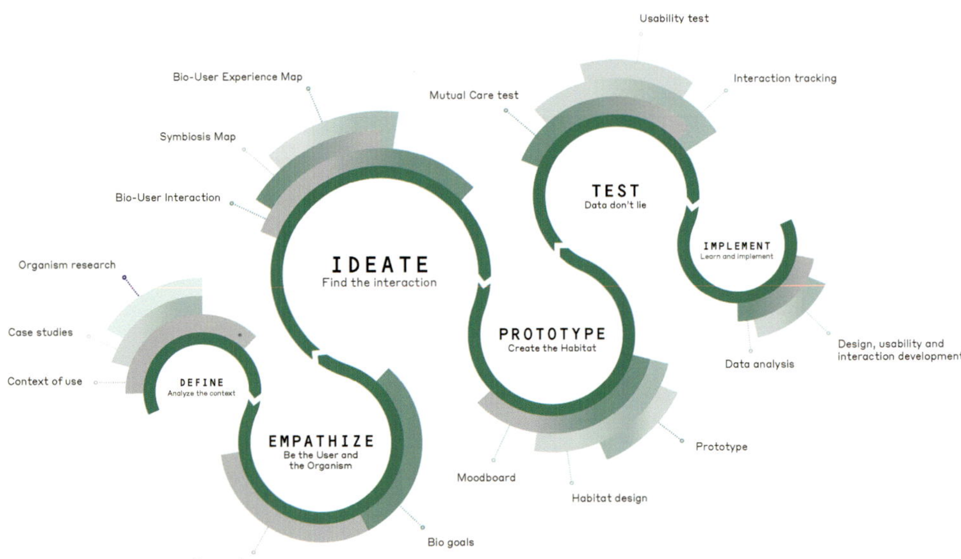

Figure 4.6 Project Flow (2024). Integration of the Bio-UX Maps within the classical Design Thinking framework, illustrating how biological and user-centered insights can be combined throughout the design process.
Credits: Elena Albergati.

support the principles of DILS (see Figure 4.6). The process is divided into six phases: *Define*, *Empathize*, *Ideate*, *Prototype*, *Test and Implement.*

To better align with the bio-UX methodology, the first two phases of the original design thinking process have been reversed, starting with *Define*. This reordering prioritizes the identification of the design challenge and the analysis of relevant case studies, allowing for a more accurate selection of the organism to be integrated before proceeding with the user and biological analysis.

- *Define*: This phase involves an in-depth investigation of the design context, the intended use and the specific problem to be addressed. Existing case studies are reviewed, and a preliminary selection of a suitable organism is made based on relevance and compatibility with the design objectives.
- *Emphasize*: In this step, the User Goals Map and Bio Factors Map are developed to explore human users' and organisms' needs, behaviors and characteristics. This dual analysis provides the comprehensive understanding needed to design meaningful interactions.
- *Ideate*: Here, the focus shifts to designing the human–organism interaction in a functional, ethical and mutually beneficial way. Tools such as the Bio-User Interaction Map, Symbiosis Map and traditional experience maps are used to guide ideation.
- *Prototype*: The product – often conceptualized as a habitat for the organism – is developed with a defined interaction strategy. This includes sketching, material experimentation and building prototypes using established UX and interaction design tools. Mood boards and mock-ups can support visual and functional exploration.
- *Test*: During this phase, the prototype is evaluated through testing with human users and living organisms to assess what functions effectively and where refinements are needed. This stage is crucial, as it translates earlier research and planning

into practice. Evaluation focuses on usability, behavioral responses, and the quality of the human–organism relationship. A range of tools may be employed, including usability testing, interaction tracking and the Mutual Care Test, which assesses whether the symbiosis envisioned in the design process is achieved in practice.

- *Implement*: The final phase involves synthesizing the findings from testing to refine the design. Improvements focus on enhancing usability, interaction quality and the long-term viability of the human–organism–environment system. The product is iterated and finalized for deployment or further development.

APPLYING BIO-UX TO DILS: A CASE STUDY ANALYSIS

This section analyses a case study introduced in Chapter 3 – *Growing Light*[11] To demonstrate how the Bio-UX methodology can be applied to the design of Designed Interactive Living Systems (DILS). The application of the User Goals Map to the Growing Light project provides an overview of the user's roles, needs, motivations, and behaviours (see Figure 4.7). Drawing on insights from existing literature, the analysis shows how each stage of the Bio-UX process can be used to examine and shape the complex interactions between human users, organisms, and products. The case study shows how Bio-UX tools can facilitate the design of ethical, responsive and regenerative DILS by enabling a detailed understanding of relationships between species and how systems behave.

CASE STUDY / GROWING LIGHT

Glowing lamps powered by bioluminescent micro-algae activated by the user's movement

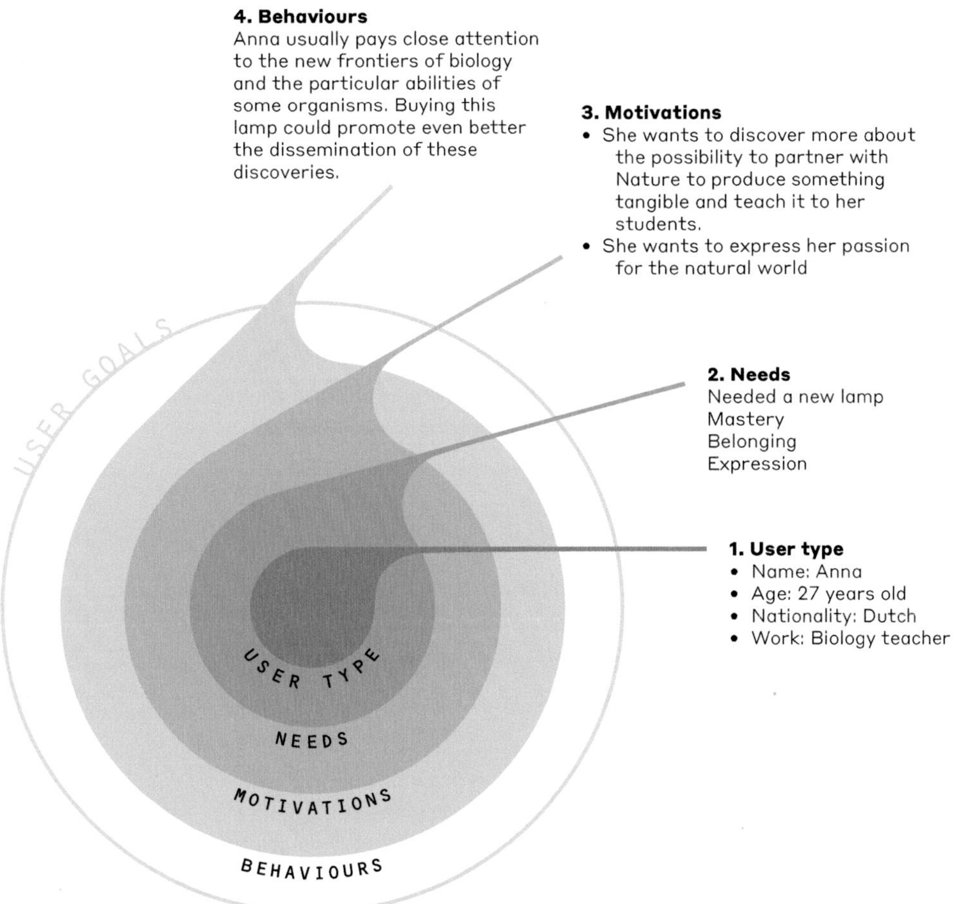

Figure 4.7 User Goals Map applied to the *Growing Light* case study, illustrating the analysis of user types, needs, motivations, and behaviours within the Bio-UX framework.
Credits: Elena Albergati.

1st STEP: USER GOALS

The first phase focuses on the analysis of the human user, supported by the User Goals Map. This tool begins with the selection of a representative target user – called a persona – whose characteristics, needs and motivations are explored to inform design decisions.

In this case, the persona is Sara, a young biology teacher living in Denmark. While her initial goal is to find a new lamp, a deeper analysis reveals more complex needs: the desire to express her passion for biology, to feel recognized for it and to communicate it effectively to her students. She is curious about collaborating with nature to create tangible, educational experiences.

Through her ongoing interest in scientific discovery, Sara learns about the bioluminescent properties of dinoflagellates and is intrigued by their potential as an educational tool and source of wonder. Motivated, she decided to purchase *Growing Light* to share this natural phenomenon with friends and students meaningfully and engagingly.

2nd STEP: BIO FACTORS

Organism*: Pyrocystis fusiformis and Lunula*

The next step is to analyze the biological characteristics of the organisms involved.

Pyrocystis fusiformis and *Pyrocystis lunula* are tropical, non-motile, epipelagic marine dinoflagellates, each up to 1 mm in length. These microorganisms emit bioluminescence when mechanically stimulated or disturbed. Autotrophic organisms derive energy from photosynthesis during the day, while their bioluminescent response is predominantly expressed at night, regulated by an internal circadian rhythm.

They require a daily cycle of 12 hours of complete darkness to ensure their viability and sustained bioluminescent activity. Their ideal living conditions include a water temperature of 18–20°C, a salinity of 3.5% and a pH range of 8.0–8.2. The medium must also be free of contaminants and supplemented with essential nutrients such as trace metals and vitamins (Sweeney, 2004). These biological parameters are summarised in the Bio Factors

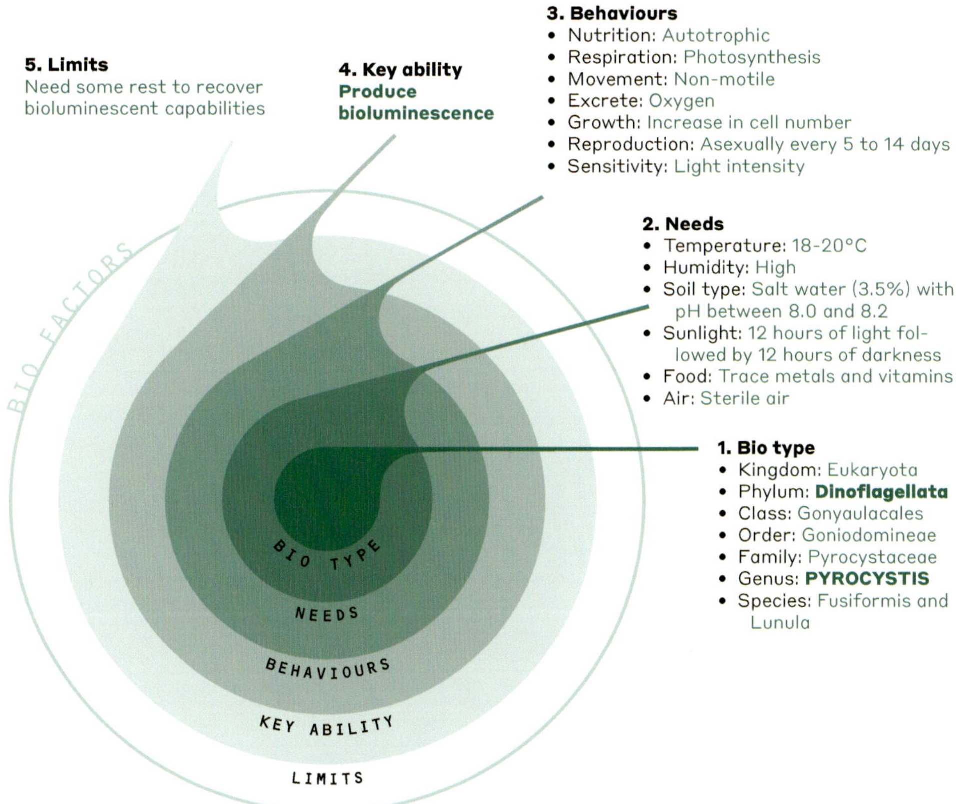

Figure 4.8 Bio Factors Map applied to the *Growing Light* case study, illustrating the analysis of the organism's biotype, needs, behaviours, key capabilities, and limits within the Bio-UX framework.
Credits: Elena Albergati.

Map (see Figure 4.8), which outlines the organism's taxonomy, environmental needs, behaviours, key abilities and limitations within the Growing Light project.

A key limitation to consider is their sensitivity to overstimulation: when subjected to continuous mechanical input, organisms can become temporarily unresponsive and stop emitting light. Therefore, rest periods between stimuli are necessary to maintain their ability to produce bioluminescence (Barati et al., 2021).

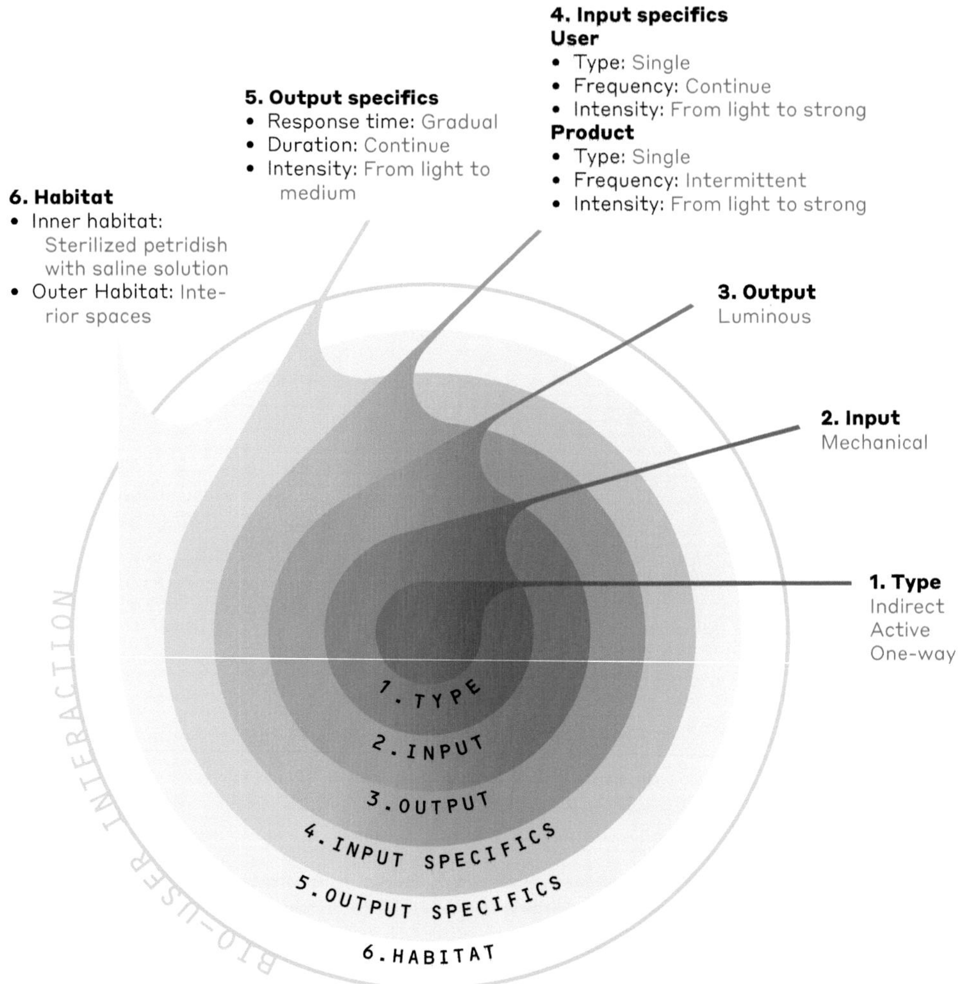

Figure 4.9 Bio-User Interaction Map applied to the *Growing Light* case study, illustrating the analysis of interaction dynamics between the human user and the bioluminescent organism across input, output, and habitat dimensions.
Credits: Elena Albergati.

3rd STEP: BIO–USER INTERACTION

This phase examines the interaction dynamics between the human user, the organism and the product. In *Growing Light*, a proximity sensor detects the human user's movement and activates a mechanical system that rotates the container. This movement stimulates the *dinoflagellates* (*Pyrocystis fusiformis and Lunula*) and triggers their bioluminescent response (see at the Figure 4.9).

The interaction is indirect, as electronic components mediate the user's input (movement), active, as the user deliberately initiates it; and one-way, as the organism's response – light emission – does not feedback to influence the user's actions.

Both inputs are mechanical: the user's movement is the initial trigger, while the device translates it into a rotational movement. These two inputs differ in timing – continuous for the user and intermittent for the device – to allow for necessary rest periods, prevent overstimulation and preserve the algae's ability to emit light. The intensity of the input can range from mild to strong depending on the user's proximity, but only a single movement is required to activate the response.

The output is a light signal with an immediate onset and short duration, in keeping with the bioluminescent behavior of dinoflagellates.

In terms of habitat, the inner habitat consists of two sterilized petri dishes sealed with a breathable membrane that allows gas exchange but prevents contamination. These are filled with a 3.5% salt solution enriched with essential vitamins and trace elements. The outer habitat corresponds to the indoor environment in which the product will be used and provides controlled lighting conditions that support the circadian rhythm of the microorganisms (Barati et al., 2021).

4th STEP: SYMBIOSIS MAP

Having mapped the needs of the human user and the bioluminescent organism and defined their modes of interaction, the final step explores the formation of a reciprocal and caring relationship. This phase examines how the user's actions – such as maintaining

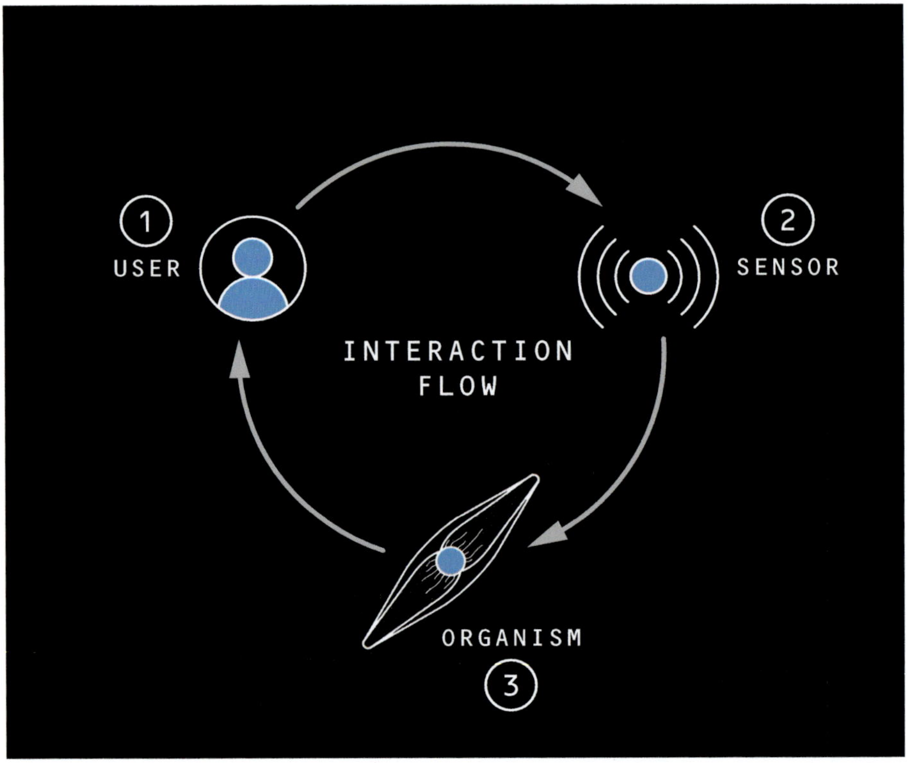

Figure 4.10 **Interaction flow between the user, the sensors, and the bioluminescent algae in the *Growing Light* project, illustrating the sequence of actions and responses that define the indirect, one-way interaction model.**
Credits: Elena Albergati.

optimal environmental conditions and respecting the organism's rest cycles – support the organism's well-being. In turn, the organism's bioluminescent response acts as feedback, potentially influencing the human user's behavior, awareness or emotional involvement. This interaction flow can be visualized in the diagram below, which maps the exchange between user input, sensor activation, and the organism's luminous output (see Figure 4.10). This mutual influence creates a loop of care, fostering a symbiotic relationship at the heart of the DILS. The Symbiosis Map illustrates the cyclical relationship between care, biological response, and behavioral change (see Figure 4.11).

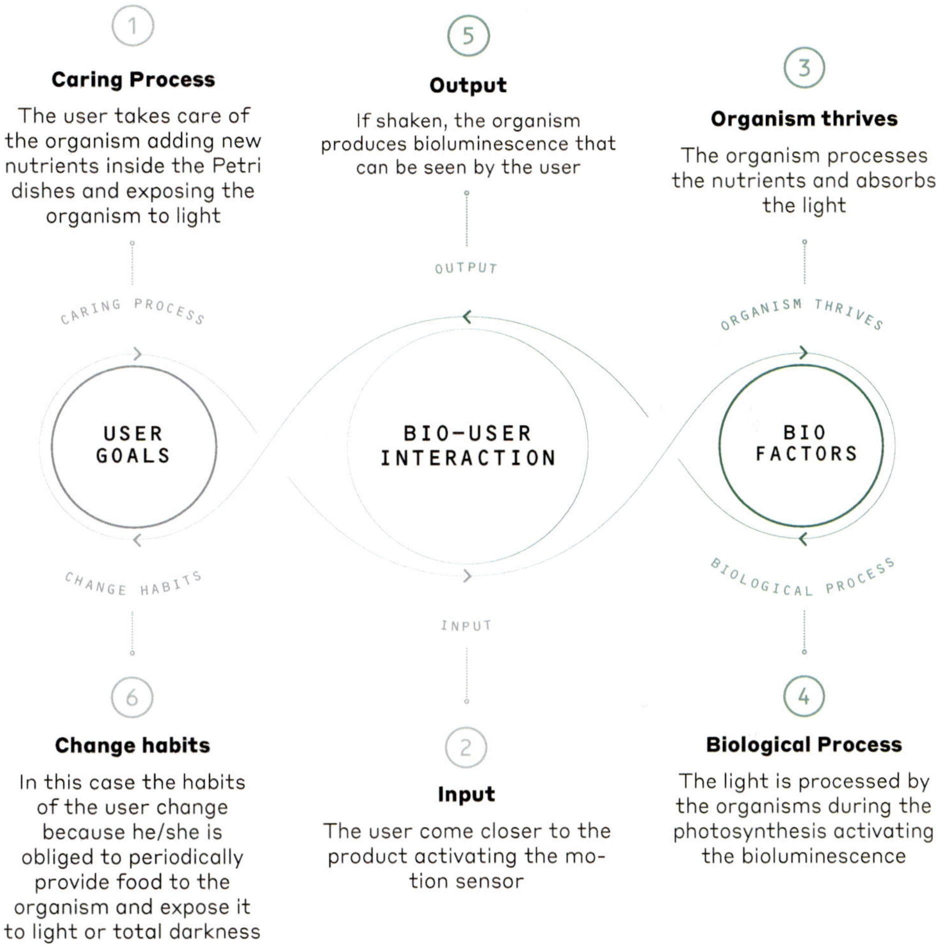

Figure 4.11 **Symbiosis Map applied to the *Growing Light* case study, illustrating the cycle of mutual care between the human user and the bioluminescent organism—from nurturing and input to biological response and behavioral adaptation.**
Credits: Elena Albergati.

This chapter has contributed to the broader framework of living interaction design by proposing a methodological approach to support the integration of biological systems into interactive design. Grounded in principles of interdependence, care, and regenerative thinking, this methodology expands traditional user-centered paradigms by incorporating non-human living agents as active participants in interaction.

Through a combination of conceptual framing and practical tools – including the User Goals, Bio Factors, Bio-User Interaction, Symbiosis, and Project Flow maps – this design process helps navigate the complexities of DILS. These tools are not intended to replace existing UX practices but to extend them toward more inclusive and ecologically embedded modes of interaction.

Reframing interaction as a multispecies co-creative process enables designers to move beyond anthropocentric logic toward symbiotic, responsive and life-enabling design practices. In doing so, it offers a pathway for operationalizing the principles of living interaction design and supports the transition toward regenerative ecologies.

NOTES

1 User Experience (UX) Design: https://www.interaction-design.org/literature/topics/ux-design
2 Yale University's Usability & Web Accessibility site offers a concise overview of user journey maps, emphasizing their role in visualizing users' step-by-step interactions to identify opportunities for UX improvements. Retrieved from: https://usability.yale.edu/understanding-your-user/user-journey-maps
3 Nielsen Norman Group provides a comprehensive "UX Mapping Cheat Sheet" that explains different types of UX mapping methods, including their purposes and contexts of use within the design process. Retrieved from: https://www.nngroup.com/articles/ux-mapping-cheat-sheet/
4 Nielsen Norman Group provides a comprehensive "UX Mapping Cheat Sheet" that explains different types of UX mapping methods, including their purposes and contexts of use within the design process. Retrieved from: https://www.nngroup.com/articles/ux-mapping-cheat-sheet/
5 Nielsen Norman Group's article on storyboarding highlights how visual narratives are used to illustrate user interactions, emotions and context over time, aiding communication and empathy in UX design. Retrieved from: https://www.nngroup.com/articles/storyboards-visualize-ideas/

6 *The Secret Sounds of Spores* is an installation created by Yann Seznec (2010) in collaboration with mycologist Patrick Hickey. As the title suggests, Seznec's intention was to make audible the movement of spores, a phenomenon that typically remains hidden in the undergrowth and thus beyond human perception. https://www.yannseznec.com/works/spores/

7 *Mould Rush* is a game based on real-time microbial growth, by Raphael Kim (2018). It is an online multiplayer game, played over several days. As the microbes grow and propagate, their growth patterns are live-streamed onto the Mould Rush Game Channel on Twitch.tv, where the gameplay takes place. https://mat.qmul.ac.uk/students/r-s-kim

8 Designed by Julian Melchiorri (2017), *Exhale* is a living bionic chandelier which purifies the air indoors. This piece explores how advances in biotechnology and engineering can be applied to everyday objects and architecture to increase the quality of our lives. The chandelier purifies the air indoors through photosynthesis performed by living photosynthetic microorganisms enclosed into leaf modules. https://www.julianmelchiorri.com/Bionic-Chandelier

9 The *Living Light Dress* was a living garment – the third in the series of bioluminescent pieces produced by Dr Victoria Louise Geaney Dr Bernardo Pollak and Dr Anton Kan (University of Cambridge). We worked with *Photobacterium kishitanni* on a nutrient-enriched agar and raw wool fabric. https://www.victoriageaney.co.uk/portfolio/living-light-dress

10 *E. chromi:* Living Colour from Bacteria is an early collaboration between artists/designers and scientists in the emerging field of synthetic biology. Project by James and Alexandra Daisy Ginsberg with iGEM 2009 Team University of Cambridge UK. https://www.daisyginsberg.com/work/echromi-living-colour-from-bacteria

11 *Growing Light:* Designing with Bioluminescent Algae by Barati et al. (2021). This project aims to provide designers first insights in the behavior of bioluminescent algae by looking at the relation between movement and light. https://repository.tudelft.nl/record/uuid:14fdd5ac-e077-403c-992a-d40a526482cf https://www.highlightdelft.nl/program/in2021/growing-light/

REFERENCES

Albergati, E. (2020). *Living interaction design: From user-centred to bio-user system* [Master's thesis, Politecnico di Milano, School of Design].

Barati, B., Karana, E., Pont, S., & Van Dortmont, T. (2021, June). Living light interfaces–An exploration of bioluminescence aesthetics. In *Proceedings of designing interactive systems conference 2021* (pp. 1215–1229). ACM. https://doi.org/10.1145/3461778.3462038

Bell, F., Woytuk, N. C., Søndergaard, M. L. J., & Alistar, M. (2024). Biodegradation as More-Than-Human Unmaking. *ACM Transactions on Computer-Human Interaction*, *31*(1), 1–41. https://doi.org/10.1145/3685526

Bødker, S., & Klokmose, C. N. (2019). The human–artifact model: An activity theoretical approach to artefact ecologies. *Human-Computer Interaction*, *34*(2), 101–138. https://doi.org/10.1080/07370024.2011.626709

Braidotti, R. (2013). *The posthuman*. Polity Press.

Campbell, N. A., & Reece, J. B. (2005). Biology (7th ed.). Pearson Education.

Garrett, J. J. (2011). *The elements of user experience: User-centered design for the web and beyond*. New Riders.

Gibbons, S. (2016). "Design Thinking 101." Nielsen Norman Group. Retrieved from https://www.nngroup.com/articles/design-thinking/.

Gough, P., Pschetz, L., Ahmadpour, N., Hepburn, L.-A., Cooper, C., Ramirez-Figueroa, C., & Catts, O. (2020). The nature of biodesigned systems: Directions for HCI. In *Companion publication of the 2020 ACM Designing Interactive Systems conference (DIS' 20 Companion)* (pp. 389–392). Association for Computing Machinery. https://doi.org/10.1145/3393914.3395908

Endmann, A., & Keßner, D. (2016). User journey mapping – A method in user experience design. *i-com, 15*(1), 105–110. https://doi.org/10.1515/icom-2016-0010

Ferreira, B., Silva, W., Oliveira, E., & Conte, T. (2015). Designing personas with empathy map. In Proceedings of the 27th international conference on Software Engineering and Knowledge Engineering (SEKE 2015) (Vol. 152, pp. 501–505). KSI Research Inc and Knowledge Systems Institute Graduate School.

Hassenzahl, M., Diefenbach, S., & Göritz, A. (2010). Needs, affect, and interactive products – Facets of user experience. *Interacting with Computers, 22*(5), 353–362. https://doi.org/10.1016/j.intcom.2010.04.002

Karana, E., Barati, B., & Giaccardi, E. (2020). Living artefacts: Conceptualizing livingness as a material quality in everyday artefacts. *International Journal of Design, 14*(3), 37–52. https://doi.org/10.1017/btd.2023.10

Kawade, Y. (2009). On the Nature of the Subjectivity of Living Things. *Biosemiotics, 2*, 205–220. https://doi.org/10.1007/s12304-009-9041-9

King, S., & Chang, K. (2016). *Understanding industrial design: Principles for UX and interaction design* (1st ed., Kindle ed.). O'Reilly Media.

Margulis, L., & Chapman, M. J. (2009). *Kingdoms and domains: An illustrated guide to the phyla of life on Earth*. Academic Press.

Moggridge, B. (2007). *Designing interactions*. MIT Press.

Norman, D., & Nielsen, J. (2002). *The definition of user experience (UX).* Nielsen Norman Group.

Norman, D. (2013). *The design of everyday things: Revised and expanded edition.* Basic Books.

Pataranutaporn, P., Ingalls, T., & Finn, E. (2018). Biological HCI: Towards integrative interfaces between people, computers, and biological materials. In *Proceedings of the 2018 CHI conference on human factors in computing systems* (Paper LBW579, pp. 1–6). ACM. https://doi.org/10.1145/3170427.3188662

Pataranutaporn, P., Vujic, A., Kong, D. S., Maes, P., & Sra, M. (2020, March). Living bits: Opportunities and challenges for integrating living microorganisms in human-computer interaction. In *Proceedings of the augmented humans international conference* (pp. 1–12). ACM.

Rafael, S., Santiago, E., Rebelo, F., Noriega, P., & Vilar, E. (2022). Bio-centred interaction design: A new paradigm for human-system interaction. In M. M. Soares, E. Rosenzweig, & A. Marcus (Eds.), *Design, user experience, and usability: Design thinking and practice in contemporary and emerging technologies* (Vol. 13323, pp. 72–87). Springer. https://doi.org/10.1007/978-3-031-05906-3_6

Rosén, A. P., Normark, M., & Wiberg, M. (2022). Towards more-than-human-centred design: Learning from gardening. *International Journal of Design, 16*(3), 21–36. https://doi.org/10.57698/v16i3.02

Saffer, D. (2009). *Designing for interaction: Creating innovative applications and devices* (2nd ed.). New Riders.

Sweeney, B. M. (2004). Bioluminescence and circadian rhythms in dinoflagellates. In T. Hastings, L. J. Kricka, & P. Stanley (Eds.), *Bioluminescence and Chemiluminescence: Progress and Perspectives* (pp. 255–258). World Scientific Publishing.

Szabo, P. W. (2017). *User experience mapping.* Packt Publishing Ltd.

Truong, K. N., Hayes, G. R., & Abowd, G. D. (2006, June). Storyboarding: An empirical determination of best practices and effective guidelines. In *Proceedings of the 6th conference on designing interactive systems* (pp. 12–21). ACM. https://doi.org/10.1145/1142405.1142410

Willis, A. M. (2006). Ontological designing. *Design Philosophy Papers, 4*(2), 69–92. https://doi.org/10.2752/144871306X13966268131514

Zhou, J., Barati, B., Giaccardi, E., & Karana, E. (2022). Habitabilities of living artefacts: A taxonomy of digital tools for biodesign. *International Journal of Design, 16*(2), 57–73. https://doi.org/10.57698/v16i2.05

CHAPTER 5

BIOLABS

CROSSROADS OF DESIGN, SCIENCE, COMPUTING AND SUSTAINABILITY FOR DILS

DOI: 10.4324/9781003366799-6

CULTIVATING SYNERGIES: WHY TRANSDISCIPLINARY BIOLABS MATTER

The transition from interactive devices made from non-renewable resources, such as petroleum-based materials or those derived from strategic and rare raw materials, to Digital Interactive Living Systems (DILS) is a significant challenge. This shift requires a transdisciplinary approach involving designers and a wide range of STEM professionals. Depending on the specific nature of the project, collaboration may require expertise from several fields, including microbiology, materials science, electronics, computer engineering, mechanical engineering and chemistry, among others. To effectively develop novel interactive artifacts and achieve meaningful experimental results, it is essential to share hybrid workspaces where design culture, life sciences and emerging technologies can coexist and interact without hierarchical dominance or mutual suspicion. These hybrid laboratories aim to foster collaboration by integrating diverse expertise (Oxman, 2016). Such facilities may be called biolabs, DIY biolabs (You et al., 2021), or SYN biolabs, the latter mainly when focused on synthetic biology. Notably, the crossing of disciplinary boundaries within these labs should not be seen as diminishing the identity or complexity of individual disciplines but rather as an opportunity to enrich and broaden the perspectives of all contributors.

Many existing biolabs are located within academic institutions, and a growing number have recently been established in design schools worldwide. This trend underlines an ever-increasing interest in biodesign research and biodesign education – a topic of contemporary relevance and active debate among design scholars. These laboratories serve a dual purpose: they provide spaces for the investigation of emerging issues, such as living systems, and offer design students the opportunity to develop both practical and theoretical skills in this rapidly evolving field. They also enable students to explore innovative forms of transdisciplinary collaboration and reflect critically on design's role in the context of living systems. In particular, biodesign education goes beyond technical skills transfer by challenging user-centered design approaches and encouraging critical engagement with the discipline's ethical, social and environmental implications (Gabrielian, 2024; Raicu & Green, 2016).

An essential contribution to this field is made by Julia Ihls and Barbara Pollini, whose research explores the role of laboratory spaces in biodesign practice and education. Their study maps a variety of biolabs, ranging from low-tech community spaces to high-tech academic institutions, and analyzes how different configurations support transdisciplinary collaborations between design and science (Ihls & Pollini, 2024; Ihls & Pollini, 2025). Ihls and Pollini highlight that the hybrid nature of biodesign often requires adaptable workspaces that deviate from traditional scientific laboratories and incorporate tools and methods not typically associated with laboratory environments. Their work highlights the potential of biolabs as critical spaces for experimentation, knowledge production and the development of new socio-technical constructs essential for the future of biodesign education and practice.

This chapter examines emerging laboratories characterized by "material tinkering" (Parisi et al., 2017) and "biotinkering" practices (Rognoli et al., 2021), focusing on the development of living materials, particularly for interaction design applications. Adopting an iterative and hands-on methodology, biotinkering plays a crucial role in advancing projects that transcend traditional design boundaries, paving the way for novel interaction paradigms based on a more-than-human perspective. Although biodesign is gaining momentum worldwide, it is still relatively rare to encounter design projects involving living organisms that successfully advance applications in fields such as human–computer interaction (HCI) or product design. The "living factor", which inherently includes characteristics such as temporality, transience and non-linearity, underlines the speculative and emergent nature of biodesign as a discipline, requiring further exploration in both research and application. In addition, several inherent challenges hinder the widespread adoption of DILS. First, integrating living organisms into interactive systems requires a deep understanding of biological dynamics and their interactions with technological frameworks. Second, managing and sustaining living organisms in interactive environments requires innovative solutions that have yet to be fully explored. Finally, the limited availability of laboratories explicitly dedicated to the intersection of biology and design – be it fashion, interaction or product design – significantly limits opportunities for experimentation and research.

Our research for this book found that very few laboratories are explicitly dedicated to this unique intersection. Nevertheless, many recent and ongoing research projects are investigating labs as essential spaces for advancing the biodesign discipline and developing new biobased and living materials (Sawa, 2016; de Lange et al., 2022; Willet, 2021; Crawford, 2023; Chappell et al., 2023; Ihls & Pollini, 2024). These biolabs are emerging as critical environments where designers can access the tools and resources needed to address material design challenges, experiment with biodesign methodologies and ultimately advance the development of DILS.

The growing interest in this domain suggests that we are only beginning to unlock the potential of these technologies. Increased investment and transdisciplinary collaboration could significantly accelerate progress in this area. Therefore, It is essential to establish laboratories that provide researchers and students with adequate facilities for experimentation and collaboration, particularly for integrating living organisms into design processes. This chapter identifies and briefly describes several established laboratories and research centers dedicated to studying and testing living materials for design applications, focusing on research involving biological interactions. These laboratories facilitate the creation of innovative projects and promote the integration of different knowledge systems needed to address current technical and conceptual challenges. As a result, they serve as fertile ground for the advancement of this emerging field.

BIOLABS: ENABLING SUSTAINABLE AND REGENERATIVE DESIGN PRACTICES

In line with Bruno Latour's assertion that "access to a laboratory can lift the world" (Latour, 1983), biolabs can be considered physical spaces and real hubs of transdisciplinary innovation. These environments allow designers, creatives' and scientists to experiment with biodesign, integrating living systems into design processes. The activities in biolabs are diverse and characterized by innovative methodologies, ranging from direct experimentation with living materials to advanced interaction design. One of the central practices in biolabs is biotinkering, which involves the study and manipulation of biological materials such as algae, bacteria, fungi and biofilms. This approach facilitates the analysis and understanding of living systems' dynamic and adaptive properties, paving the way for novel design applications (Rognoli et al., 2021). At the same time, biolabs serve as fertile ground for the development of biofabricated and regenerative solutions that meet functional requirements and aim to regenerate resources and minimize environmental impact, in line with the principles of sustainability (Filippi et al., 2024; Wilson, 2023).

A key focus of biolab's activities is the design of multispecies interactions, which involves studying and creating artifacts that enable symbiotic relationships between humans and living organisms. This type of design, inspired by the principles of more-than-human design, emphasizes the interconnectedness of different species and promotes a vision of design that transcends anthropocentric paradigms (Giaccardi & Redström, 2020). In addition, biolabs use advanced tools and machinery for prototyping materials and artifacts that seamlessly integrate biological components with technological systems, fostering innovative experiments in interaction design.

Biolabs are increasingly recognized as key hubs for advancing research in human-computer interaction (HCI) and fostering the development of DILS. The intersection between biodesign and HCI has grown in recent years, merging the digital and biological domains by integrating computational technologies with biofabrication practices. As these research areas within HCI continue to evolve, scholars are beginning to observe (Bell et al., 2024a; Merritt et al., 2020; Forman et al., 2023; Dade-Robertson & Zhang, 2024) that practitioners are drawing significant inspiration from biodesign, mainly through DIY approaches that involve experimenting and tinkering with living organisms. To support these innovative practices, traditional HCI labs – once limited to computers, digital fabrication tools and electronics workstations – are now expanding to include biological wet labs, microscopes, cooking equipment and incubators. This integration of biodesign and HCI has catalyzed the emergence of novel biodigital technologies that challenge existing paradigms of computational interaction and foster a more sustainable technological future. As biodesign research continues its rapid expansion, these hybrid spaces offer unprecedented opportunities to explore new interaction models where living systems serve not as passive subjects but as active, dynamic agents within the design process. Recent studies in HCI in fact have shown that the integration of microorganisms into artifacts introduces a unique quality – liveness – leading to the creation of "living artefacts" (Zhou et al., 2022; Groutars et al., 2024; Zhou et al., 2024; Risseeuw et al., 2024; Kim et al., 2023). This integration offers groundbreaking interaction possibilities by merging biological and technological elements. Such an approach provides unprecedented opportunities for research and application of DILS, combining biological and technological innovation in a transformative way, as illustrated by the case studies presented in Chapter 3.

Thus, biolabs are sites of technical experimentation and spaces for envisioning a future in which technology co-evolves with living systems, moving towards a more inclusive and regenerative design paradigm. Transdisciplinarity is a fundamental element of these labs, as it is essential to foster synergistic collaborations between biologists, bioengineers and designers. These collaborations are crucial for translating speculative projects into tangible, productive realities, bridging the gap between academic research and applied innovation. In this context, the concept of "biofactories" (biofoundries) is particularly relevant. Guided by the principles of synthetic biology, biofactories have the potential to produce biological artifacts, biomaterials and biofuels, thereby revolutionizing production processes (Hillson et al., 2019; Holowko et al., 2021).

Although the number of laboratories dedicated exclusively to the intersection of living biology and design remains limited, biolabs are increasingly emerging as essential platforms for exploring the potential of biodesign and unlocking new design possibilities. Scholars conceptualize these laboratories as focal points for integrating synthetic biology and critical design, enabling the creation of hybrid artifacts and innovative product-service systems that carry unique meanings, interactions and narratives (Langella, 2024). From a design perspective, biolabs are more than spaces for creation; they are vital arenas for learning and training. Their experimental nature fosters pedagogical innovation, enabling students and researchers to learn directly and hands-on. Biolabs equip individuals to develop sustainable and regenerative solutions by combining biology, engineering and design expertise. In addition, designers play a critical role in shaping the cultural and ethical values needed to ensure societal acceptance and market viability. In biodesign, critical and speculative design often accompanies experimentation, providing a framework for exploring the broader implications of biological and technological innovations (Ginsberg & Chieza, 2018; Ginsberg et al., 2017).

Yet despite their potential to drive innovation, designers often lack the scientific literacy to navigate the complexities of the life sciences effectively. Designers must acquire the knowledge and skills necessary to collaborate with scientists and bioengineers, ensuring scientific rigor and ethical awareness. This underlines the urgency of establishing biolabs dedicated to biodesign education, where students can cultivate interdisciplinary skills through direct, experiential learning.

In selecting laboratories for this chapter, we have prioritized those focusing on projects related to living interaction. As the field is still emerging and lacks an extensive network of dedicated laboratories, most facilities highlighted here are biolabs where design experiments with living organisms are actively conducted. As the biolab landscape is expected to evolve and expand significantly in the coming years (Ihls & Pollini, 2024), this selection is not intended to be exhaustive but rather to highlight currently active biolabs where design plays a central role.

BIOLABS: HUBS FOR BIODESIGN AND INTERACTION DESIGN

Biodesign is developing rapidly worldwide, but most well-established biolabs are predominantly in Western countries, particularly in regions where universities, research institutions and organizations specializing in bioinformatics, synthetic biology and HCI are more advanced. Some notable biolabs involving designers are located within universities and art schools, which are increasingly opening their doors to new practices in biodesign and biological interaction design.

Below, we present a selection of labs and research initiatives from different countries, highlighting not only those focused on HCI but also promising institutions advancing biodesign through work with living materials and broader research areas. These initiatives may have significant potential to contribute to HCI applications in the future.

Let's start with one of the first established biolabs in the United States, founded at the Rhode Island School of Design by RISD faculty member Edna Lawrence in the early twentieth century. The *Edna Lawrence Nature Lab*[1] was designed to awaken students to the awe-inspiring beauty inherent in nature's designs, shapes, colors, textures and structures. It continues its tradition of providing unrestricted access to authentic natural history specimens while encouraging creative exploration in the climate crisis and biodesign. There is a dedicated biodesign makerspace where students are invited to interact with elements of nature through activities such as culturing organisms for material creation, creating metabolisms in simple chemistry-type set-ups, making shapes that mimic organismal design from simple 3D materials and creating microcosms and ecosystem set-ups using fish tanks, pumps and other wet-lab, grow-lab–type equipment. Students interact with living organisms for educational purposes to deepen their understanding of "natural technologies" and how nature designs itself, and they use this as a springboard for designing colors, textures and structures. The lab continues its tradition of providing unrestricted access to authentic natural history specimens while encouraging creative exploration in the climate crisis and biodesign. There is a dedicated biodesign makerspace where students are invited to interact with elements of nature through activities such as culturing organisms for material creation, creating metabolisms in simple chemistry-type set-ups, making shapes that mimic organismal design from simple 3D materials, and creating microcosms and ecosystem set-ups using fish tanks, pumps and other wet-lab, grow-lab type equipment. Students interact with living organisms for educational purposes to deepen their understanding of 'natural technologies' and how nature designs itself, and they use this as a springboard for human design.

The *MIT Media Lab*,[2] is undoubtedly one of the most renowned research facilities in the United States. It hosts several research groups that specialise in biological HCI frameworks. The Tangible Media Group[4] led by Professor Hiroshi Ishii, and the Fluid Interfaces Group.[5] are two such groups that explore interactive experiences between users and biodigital systems, focusing on wearable biotechnology and growable interfaces.[3]

In particular, the Tangible Media Group investigates the concepts of Tangible Bits and Radical Atoms, which aim to bridge the gap between the digital and physical worlds by giving data a dynamic material form and enhancing multisensory human–computer interaction.

Building on these contributions, the 2023 CHI Conference – the leading venue for human–computer interaction research – hosted a dedicated workshop titled 'Living Bits and Radical Aminos', gathering interdisciplinary researchers to discuss new biodigital interfaces and applications (Forman et al., 2023).

Beyond the MIT Media Lab, other institutions have also played a crucial role in advancing research into living interaction design. One such institution is the *Center for Science and the Imagination* at Arizona State University[6] which has developed an influential framework based on three key elements: biological materials, intermediate platforms and user interaction (Pataranutaporn et al., 2018). Biological materials include DNA, RNA, enzymes, genes, epigenetic information, mutation, reproduction, self-assembly and biological forms. These materials allow designers to manipulate entities at different biological scales – much as digital materials are manipulated in traditional HCI contexts.

Intermediate platforms act as bridges, connecting biological materials with users or digital systems and enabling the seamless exchange of information between them. This integrative framework facilitates the development of biodigital systems that combine biological and computational processes, unlocking new opportunities for interaction, design and innovation.

At the University of Colorado Boulder, within the ATLAS Institute and the Department of Computer Science, the *Living Matter Lab*[7] is led by Mirela Alistar.[8] The lab was established to use emerging technologies to empower individuals by providing insights into their biology and biome, ultimately making healthcare more affordable, effective and personalized. It combines expertise from multiple disciplines, including computer science, nanotechnology, biology, microbiology, genetics, social sciences and design. One of its key tools is the digital microfluidic biochip, which enables direct human interaction with the microbiome for diagnostic and therapeutic purposes.

In their article "Making Biomaterials for Sustainable Tangible Interfaces", (Bell et al., 2024b), Bell and the co-authors describe a studio designed to provide a platform for HCI practitioners to experiment with creating tangible interfaces using eco-friendly biomaterials. These include bioplastics – primarily derived from agar, alginate and gelatine ingredients – and bioclays made from compost, cornmeal and eggshells.

Another notable project from the Living Matter Lab, titled *Living Interfaces: Human Microbiome* (Bell et al., 2023), explores the intimate relationship between humans and the invisible ecosystem of microbes that live in symbiosis with the human body. This project introduces a microbiological dimension to HCI by exploring the skin microbiome as a unique and intimate material for interaction design, positioning the human body itself as a microbial interface.

The *BioInteractive Design (BInD) Lab*[9] at Northeastern University develops biomediated materials and systems to improve human and environmental health. Located in the Interdisciplinary Science & Engineering Complex (ISEC) – a major hub for interdisciplinary collaboration – the BInD Lab operates as a state-of-the-art Biosafety Level 1 laboratory and design studio that brings scientists, designers and engineers together to drive innovation through interdisciplinary research.

The BInD Lab integrates synthetic biology, design and engineering to create systems and materials that incorporate desirable biological functions into human products and the built environment. These include applications for health diagnostics, pathogen detection, toxin remediation, stress mitigation, waste processing, energy generation and biological communication. The laboratory enables the seamless integration of biological processes into functional materials and interactive systems by developing advanced manufacturing platforms for embedding synthetic DNA gene circuits.

Beyond its materials innovations, the BInD Lab also explores the interface between human and ecological systems by creating visualizations and interactive experiences. These projects aim to reveal natural interactions – such as microbial ecologies, circular resource flows and multispecies relationships – by making them visible, tangible and interactive. The lab promotes a deeper understanding of ecological dynamics and encourages sustainable behaviors rooted in systems thinking and biocentric design.

At the University of Maryland, Baltimore County (UMBC), the *Designing pARticipatory futurEs (DARE) Lab*,[10] led by Dr Foad Hamidi within the Department of Information Systems, focuses on HCI research with an emphasis on participatory design. The lab specializes in the participatory design and evaluation of emerging systems, including digital living media systems and personalized interfaces tailored to diverse user groups.

In the area of living media interfaces, the DARE Lab has developed *Rafigh: A Living Media Interface for Children*, designed to support speech and language intervention for children with disabilities. This innovative system fosters an empathic relationship between the child and a living mushroom colony, creating a unique therapeutic interaction (Hamidi et al., 2014).

Recent projects include *Space for All*, a collaboration with Digital Harbor Foundation and MN Associates, to create accessible and engaging learning spaces and programs for youth and young adults with autism in Baltimore City.[11] Another initiative, *Introducing Synthetic Biology using Biomaker Activities*, engages youth and their families in co-designing activities to support STEM learning and engagement in collaboration with the University of Texas at El Paso, the Baltimore Underground Science Space (BUGSS), and MN Associates.[12]

The DARE Lab also serves as a collaborative space for building physical prototypes, equipped with tools such as laser cutters, 3D printers, 3D scanners, vinyl cutters, soldering stations and microcontrollers.

Through these diverse projects and resources, the DARE Lab continues to advance the field of HCI by integrating participatory design principles with emerging technologies to address the needs of diverse communities.

The *INCUBATOR Art Lab*[13] at the University of Windsor is a hybrid BSL-2 laboratory and creative space at the intersection of art, science and ecology. Founded in 2009 by Jennifer Willet, it supports bioart research, education, exhibitions, and multimedia performance using both scientific and DIY tools. The lab functions as a site for experimenting with biotechnologies as embodied performative practices and promotes accessibility and public engagement through workshops, installations, and live events.

Moving from the US to Europe, the *BioLab at the Burg Giebichenstein University of Art and Design Halle* is a dynamic center for developing innovative, forward-looking designs that combine art, design and biotechnology. The lab critically examines current biotechnological paradigms and offers fresh perspectives rooted in artistic and design methodologies. A key focus of BioLab's research is the exploration of hidden life forms through design approaches, intending to reveal previously unexamined aspects of biology and technology. The lab emphasizes the study of living manufacturing processes, renewable resources and the development of adaptive, resilient products and systems. These areas are seen as essential components of a forward-looking approach to design that promotes sustainability, creativity and innovation.[14]

Recent projects at the BioLab include *Living Layers*, which explores the embedding of living microorganisms in materials, membranes and textiles by combining textile technologies with biotechnological processes. This project explores the potential of symbiotic systems to create surfaces and structures with novel properties, assessing the role of living organisms in material design and their interactions with the environment.

Another notable initiative is *Bio.Lumina*, which focuses on the biochemical and structural properties required to sustain microbial activity in new materials. The study focuses on bioluminescent bacteria such as *Aliivibrio fischeri*, which have symbiotic relationships with marine organisms such as the Hawaiian bobtail squid (*Euprymna scolopes*). The project aims to integrate living microorganisms into products, creating active microbial materials suitable for manufacturing techniques such as 3D printing and electrospinning.

The BioLab is equipped with state-of-the-art bioscience equipment, providing organisms and chemicals required for common bioscience methods. Students have the opportunity to experiment with microorganisms such as fungi, bacteria, and algae and their products, including fibers, bioplastics, cellulose and dyes. The laboratory fosters a culture of openness and transparency, facilitating objective discussion and exploration at the interface of nature, culture and technology.[15]

Located at the Hochschule für Gestaltung Karlsruhe (HfG Karlsruhe), *the Bio Design Lab*[16] is a dynamic and hybrid environment that bridges the digital and physical realms. Conceived as a platform for connection and collaboration with local partners and resources, the lab facilitates the presentation, education and transfer of knowledge in biodesign. Students, professionals, and visitors are encouraged to engage with biodesign projects exploring diverse and multidisciplinary research areas, including DILS.

Projects within the Bio Design Lab focus on the local region, its materials and its possibilities, and they aim to rethink modes of production in Karlsruhe and southern Germany. Topics under investigation include algae, soil, plants, the human body and agriculture. The lab gathers regional resources by connecting with local partners, developing a know-how glossary and a materials library to promote sustainable materials development. Virtual workshops and multimedia presentations of outstanding sustainable materials projects are used to disseminate knowledge.

The lab functions as an incubator and model for collaboration and production, which is in line with the ongoing transformation of the HfG Karlsruhe. It was first presented to the public during the "Critical Zones" exhibition at the ZKM | Centre for Art and Media.

As part of their research on the influence of spatial contexts in biodesign, Julia Ihls and Barbara Pollini are investigating the role of laboratory environments in shaping biodesign practices through the study *From Observation to Operation: How Lab Spaces Influence Biodesign Practice* (Ihls & Pollini, 2024, 2025). This research examines how different types of laboratory environments – from low-tech DIY configurations to high-tech institutional facilities

– influence both the practice and pedagogy of biodesign. The study highlights the importance of curricula and working environments in cultivating hybrid professional roles integrating design and science. It also highlights the growing need for adaptable, transdisciplinary laboratory spaces to accommodate the evolving methodologies and collaborative nature of contemporary biodesign (Pollini, 2024).

Another biolab in Germany is part of the *Matters of Activity. Image Space Material*[17] Cluster of Excellence at Humboldt University in Berlin. This interdisciplinary research initiative explores the intersections of biology, design, and materials science, with a strong focus on active materials, which are materials that respond and adapt to environmental and functional stimuli. The cluster is dedicated to rethinking materiality through the lens of activity, promoting sustainable and creative approaches in design, architecture and science.

At the heart of this initiative is a state-of-the-art biolab integrated into a larger research infrastructure that includes experimental fabrication spaces, digital tools and computational facilities. The biolab supports hands-on experimentation with living systems and bioinspired materials, enabling researchers and designers to explore natural phenomena such as self-organization, growth, filtration and adaptability and translate them into material and design innovations.

The cluster is organized into six interrelated research strands – weaving, filtering, cutting, material form function, object space agency, and symbolic material – and each explores how material activity can inform future-oriented design and production practices. Facilities available to support this work include:

- Wet and dry biolabs for growing microorganisms and processing organic matter
- Digital fabrication labs with 3D printers, CNC machines and laser cutters

- Soft robotics labs for developing responsive and active materials
- Textile labs for investigating biointegrated and programmable woven structures
- Materials analysis tools, including microscopy, spectroscopy and imaging systems
- Visualization and immersive media labs to simulate material behavior and interaction

Current research topics include soft robotics, active textiles, biobased composites and biological filtration systems, with applications ranging from architecture and healthcare to product and interaction design. Exploring bio-integrative materials is also opening up new avenues for developing more ecologically sustainable products.

The cluster also plays a vibrant cultural and public role through exhibitions, workshops and symposia, fostering dialog between science, art and society. A key example is the upcoming *_Matter*[18], which invites the public to rediscover the agency of matter in shaping everyday life. In Berlin, the festival features exhibitions, debates and hands-on workshops exploring the intersections of materials research, artistic experimentation and societal transformation. Visitors experience an adventurous journey that reveals how material activity permeates and informs the world around us.

In Belgium, *the Laboratorium*,[19] based at the Media Arts Studio of KASK/School of Arts Ghent, is an experimental laboratory that explores the intersections between art, design and biotechnology and promotes innovative approaches that integrate these disciplines. Located next to Formlab, a facility specializing in 3D additive manufacturing technologies, the Laboratory actively promotes links between biotechnology and 3D printing. This proximity encourages collaborative projects that combine biological processes with advanced manufacturing techniques, leading to novel applications in art and design.

A notable initiative within the Laboratorium is The Color Biolab project, which contextualizes color in contemporary art and design while addressing environmental concerns associated with the color industry. The project explores sustainable alternatives by producing pigments from microalgae cultivated on KASK's wastewater and introducing circular biobased economic systems within the institution. The research also explores structural coloration – producing color through microscopically structured surfaces – to innovate in art and design applications.

The Laboratory collaborates with artists and scientists to develop projects that apply research from the Color Biolab, further enriching the dialog between art, science and technology.

The Laboratorium collaborated with other partners to realize the *ETHO Online Session: New Material Practice in Biolabs.*[20] This event was hosted by ELIA, the globally recognized network for higher art education, and co-organized with institutions such as the Royal College of Art, Aalto University and the Academy of Fine Arts Vienna, among others.

The session aimed to explore how emerging material practices involving living organisms and biological processes are being integrated into art and design education. By bringing together researchers, educators and practitioners from different disciplines, the event fostered discussions on transdisciplinary approaches, sustainable practices and the evolving role of biolab infrastructures within art schools. It also highlighted both the opportunities and challenges of embedding biofabricated materials research and DILS in educational contexts.

The Centre of Design Research for Regenerative Material Ecologies (DREAM)[21] at the Faculty of Industrial Design Engineering at TU Delft represents a pioneering initiative in the integration of living materials into the design process for the creation of living artifacts.

Professor Elvin Karana leads the center to a transdisciplinary approach, bringing together design, biology, materials science and engineering to imagine new relationships between humans, organisms, materials and the environment. At the heart of DREAM is a biolab, established in 2022,[22] which will provide the infrastructure for growing and prototyping living systems. The lab enables a new generation of design research that engages directly with biofabrication, genetic engineering and organism–material interactions. It facilitates both speculative and applied experimentation, positioning living materials as active components in design processes.

A key research frontier at DREAM is redefining HCI through the lens of living artifacts. Karana and her team are investigating how biological responsiveness, growth, decay and metabolism can be integrated into interactive systems for regenerative ecologies (Karana et al., 2023) and designed into interactive systems that evolve and engage users in symbiotic and caring relationships with other living beings. In this context, biological entities are treated as input–output components and co-performers in dynamic, multispecies interaction scenarios (Groutars et al., 2024).

One of the lab's significant research streams explores how organisms such as mycelium, bacterial cellulose and cyanobacteria can act as interfaces capable of communication, adaptive behavior and even microbial agency in interactive environments. These materials are being investigated for their potential to create living artifacts, interactive therapeutic elements and regenerative wearable interfaces (Kim et al., 2024; Zhou et al., 2024). One notable example is the *NextSkins project*,[23] in which Karana's team is developing biofabricated microbial skins that respond to environmental or physical conditions, such as skin inflammation. These living skins blur the line between medical devices and living organisms, rethinking what it means to make medical devices and wearables.

The experimental approach of the designer in the biolab is also evident in projects such as *Living with Cyanobacteria*, which explores how human–microbe relationships can be cultivated through care, material agency and time (Zhou et al., 2024). In addition, tools such as FlavoMetrics and the Cyano-chromic interface explore real-time interaction with microbial color expression, advancing a taxonomy of "living aesthetics" for HCI (Risseeuw et al., 2024).

Through these initiatives, DREAM challenges conventional HCI paradigms, shifting the focus from interaction with computational systems to cohabitation with living systems. It promotes a vision of material ecologies that are both regenerative and sustainable and experientially and ethically transformative – inviting designers and users to rethink their role in the broader web of life.

The Centre of Design Research for Regenerative Material Ecologies (DREAM) was part of the *Material Experience Lab*, an international research group active from 2015 to 2025 and dedicated to exploring the intersection of design and emerging materials. It was co-founded by Professor Elvin Karana, Chair of Materials Innovation and Design at the Faculty of Industrial Design Engineering at TU Delft, and Associate Professor Valentina Rognoli from the Design Department at Politecnico di Milano. While the *Material Experience Lab*[24] has now concluded its activities, Karana and Rognoli continue to lead their respective research groups and collaborate closely in pursuit of their shared vision of advancing design and material innovation.

The Materials Design for Transition research group[26] was founded in 2021 by Valentina Rognoli and Sofia Duarte Poblete. Building on the work and vision initiated within the Material Experience Lab, the group continues to emphasise research into materials that can facilitate a transition towards more regenerative models of production and consumption. The group explores materials and biodesign as catalysts for sustainable and social change, focusing on applications across various design fields and embracing more-than-human and multispecies perspectives that recognise the agency of living and non-living matter in design processes.

In July 2025, based on this initial work, the group established the Biodesign Lab within the Design Department and School of Design at the Politecnico di Milano. The lab was created thanks to the commitment of the group's young researchers and the support of the MICS (Made in Italy Circular and Sustainable) project.[25]

Together, the Materials Design for Transition group and the Biodesign Lab are the main point of reference for biodesign and materials design research at the Politecnico di Milano, promoting an experimental, collaborative and multidisciplinary approach to regenerative, more-than-human and bioinspired material design.

Another of Italy's most well-established centres for bio-inspired and bio-based design is the Hybrid Design Lab.[27] Founded in 2006 at the Second University of Naples, it is now based at the University of Naples Federico II under the coordination of Carla Langella. The lab explores the reciprocal relationship between design and the life sciences, transforming theoretical and experimental biological research into innovative, sustainable products and services.

With a focus on biointeraction design, the lab's projects frequently utilise microorganisms such as Physarum polycephalum and bacterial cellulose to investigate novel material systems and living interfaces (Langella & Fiume, 2023). The lab's interdisciplinary research involves collaborations with technological and scientific partners, including *BIOlogic*,[28] a research initiative that develops materials derived from biological growth processes.

Over the years, the Hybrid Design Lab has become a leading hub for experimentation at the intersection of design and biology, advancing sustainable innovation and promoting the responsible integration of science into everyday life.

The *Sustainable Design Lab*[29] at the University of Florence is an interdisciplinary research center committed to addressing pressing environmental, social and political challenges through innovative design practices. Integrating different disciplines – including biodesign, circular design, eco-social design and sustainable digital design – promotes a holistic approach to sustainability that combines creativity, scientific research and social responsibility. Its transdisciplinary approach fosters collaboration between designers, biologists, engineers and social scientists, enabling the development of innovative frameworks and methodologies that redefine traditional design practices.

By combining scientific expertise, such as the contribution of the in-house biologist, with design thinking, the lab ensures that its work addresses both technical and human-centered aspects of sustainability. The lab actively engages in research, education and real-world applications, working with local and global communities to implement scalable and impactful solutions.

To return to the Netherlands, which has one of the highest concentrations of multidisciplinary spaces dedicated to biodesign inside and outside of academia, providing access to freelance designers and independent researchers, the *Material Aesthetics Lab*[30] at Eindhoven University of Technology (TU/e) stands out as a leading research hub dedicated to advancing biodesign practices centered on living systems. The lab explores the intersection of materials, aesthetics, interaction design and biology, working at the frontier of emerging relationships between humans and non-human entities. With a strong focus on sustainability, responsiveness and material intelligence, the lab promotes new forms of interaction that integrate life processes into design. A central component of the Lab is the Bio Lab (ML-I),[31] a dynamic and highly specialized environment designed for interdisciplinary biomaterials research. The Bio Lab is equipped to support the aseptic techniques required for cultivating, preparing and post-processing living and biobased material samples. It also houses two customizable and multifunctional bioprinters, enabling researchers to explore novel additive manufacturing techniques at the interface between biology and design.

The research carried out in the Material Aesthetics Lab covers a wide range of topics, including sustainable soft robotics with living or biodegradable components; more-than-human temporalities, exploring how biological and ecological rhythms reshape design; the materiality of care, examining how materials can foster affective and empathetic engagement; inclusive gastronomy, focusing on multisensory, culturally aware food experiences; and haptic experience design, creating tactile environments that respond to human and non-human touch.

The lab's facilities are designed to support crossdisciplinary experimentation and empower investigation in HCI and Designed Interactive Living Systems. In addition to the Bio Lab, the space includes advanced additive manufacturing technologies, material processing tools and precision instrumentation for measuring material behavior. The lab is equipped to maintain the aseptic conditions necessary for preparing, growing and post-processing material samples. It also pushes the boundaries of additive manufacturing by using two multifunctional and customizable bioprinters, enabling innovative explorations at the intersection of biology and design. Using the resources of the Bio Lab, the research group conducts hands-on experiments with microorganisms, fungi and other living entities to uncover new ways of integrating life into design practices. Their work significantly shapes the emerging field of biodesign by addressing critical issues such as sustainability, material aesthetics and the evolving relationships between humans and living systems. The lab operates as an open studio, encouraging collaboration between designers, engineers, biologists, chemists, physicists and artists, fostering new methods and practices that transcend disciplinary boundaries.

In addition to its academic research, the lab engages in collaborative projects with external partners – from hospitals and research institutes to design studios, restaurants and public institutions – to ensure that its outputs are innovative, socially relevant and applicable.

Through this integrated approach, the Material Aesthetics Lab is playing a crucial role in redefining contemporary design paradigms, placing livability, adaptability and care at the heart of material innovation. It envisions a future where materials and artifacts are no longer static but responsive, evolving and alive – reflecting a deeper engagement with the dynamic nature of living systems. This vision aligns with and actively contributes to the emerging field of DILS, where biological intelligence, material responsiveness and multispecies interaction converge in novel design expressions.

Material Incubator (MI)[32] is an interdisciplinary creative research lab in the Netherlands that explores collaborative processes with living (micro)organisms such as algae, fungi, bacteria and plants. Founded by Elvin Karana (Avans University of Applied Sciences) and Rens Holslag (Caradt), MI is a collaborative platform bringing together researchers from Caradt, MNEXT, the Materials Experience Lab (TU Delft) and a wider network of partners. The lab explores living materials and artifacts through design practices focusing on care, reciprocity and more-than-human storytelling.

Combining researchers, practitioners and students from *Avans University of Applied Sciences*, MI fosters tangible interactions with living organisms to understand their unique qualities and constraints. It employs diverse technical and creative methods, treating artists, designers and scientists as equal and active partners in the material creation process. The core research team includes artists and product designers specializing in biodesign, (digital) storytellers and microbiologists working together at the intersection of life sciences and material innovation.

Founded in 1994 by Marleen Stikker and Caroline Nevejan, *Waag Futurelab*[33] has established itself as a pioneering institution at the intersection of technology, art and society in Amsterdam. As a transdisciplinary hub, Waag brings designers, artists, scientists and citizens together to explore and co-create the future through public research methods. Its mission is to empower individuals and communities to collectively shape an open, inclusive and equitable society based on collaborative design and responsible innovation. At the heart of Waag's philosophy is a commitment to public research, which places citizen engagement and social context at the center of technological and societal development. This ensures that emerging solutions are technically feasible, culturally meaningful, ethically sound and socially embedded. Within the Waag structure, the *Open Wetlab*[34] emerges as a pioneering space with all the facilities for experimenting with living organisms, dedicated to biodesign, bioart and participatory biology. The lab explores the social and environmental implications of biotechnologies and provides a platform for experimental practice and critical engagement with emerging life-based innovations.

A flagship initiative of Waag's Open Wetlab is the *BioHack Academy.*[35] This annual educational program empowers participants to build and operate their own biolabs using open-source hardware, conduct hands-on biological experiments and develop a critical perspective on synthetic biology. Since its creation, the Academy has expanded across Europe, the Americas and Asia, establishing itself as a model for distributed, accessible and democratized scientific practice. As one of the first public biolabs in Europe, the Open Wetlab has pioneered the fields of biodesign and biofabrication in non-academic communities. It promotes a civic and reflexive engagement with living systems, enabling designers, artists and citizens to work directly with biological materials and processes. Programs such as the BioHack Academy equip participants with the technical skills and conceptual tools needed to navigate and co-create with the living world – skills that are increasingly essential in developing DILS.

Supported by the *Hybrid Lab Network*,[36] a European initiative connecting interdisciplinary labs in art, science and society, the Open Wetlab is embedded in a broader movement towards collaborative and transdisciplinary innovation. Through this ecosystem, the Waag Futurelab significantly contributes to the evolution of DILS by promoting a model of public research that integrates participatory design, accessible biotechnology and more-than-human interaction.

Waag's vision of citizen science, biohacking and inclusive innovation is directly aligned with the principles of DILS. These systems are not simply advanced biological technologies; they are living artifacts designed to interact dynamically with both human and non-human agents, cultivating symbiotic relationships, ecological awareness and regenerative practices. By promoting accessibility, critical reflection and open knowledge, Waag is helping to shape a future in which biotechnology is not the exclusive domain of institutions but a shared and evolving cultural practice – ethically interrogated and creatively reimagined by a diverse community of stakeholders.

BlueCity Lab[37] in Rotterdam is a pioneering community hub and biodesign incubator at the forefront of the transition to a circular and regenerative economy located in the repurposed changing rooms of the former Tropicana tropical swimming pool. Uniquely recognized as the world's first circularly built biolab, BlueCity Lab provides a dynamic platform for "bioneers" – biobased entrepreneurs, designers and researchers – to experiment, prototype and validate sustainable products and processes in line with planetary boundaries and regenerative economic principles.

BlueCity Lab supports innovation through a holistic biodesign infrastructure consisting of four specialized labs, each tailored to a specific aspect of material and biological experimentation:

- *Fabrication Lab:* A prototyping and fabrication space where circular design principles are translated into tangible, functional artifacts.
- *Biochemistry Lab:* Equipped with incubators, fume hoods, microscopes and magnetic stirrers, it supports light chemical processes and provides a flexible, safe environment for exploring chemical-biological innovations.
- *Fungi Lab:* Dedicated to the cultivation and experimentation with non-pathogenic, non-sporulating fungal cultures, facilitating the development of mycelium-based materials and processes.
- *Microbiology Lab:* Designed for working with non-pathogenic bacterial cultures and equipped with laminar flow hoods, microscopes and incubators. It provides ideal conditions for microbial research in support of innovative biobased design.

BlueCity Lab provides fertile ground for developing DILS by enabling hands-on experimentation with living materials such as fungi and bacteria. These systems involve interfaces and artifacts integrating biological responsiveness, opening up possibilities for adaptive, symbiotic and evolving relationships between human and non-human agents. For example, microbial and fungal properties explored in the Microbiology and Fungi Labs could inspire interactive systems that sense, respond or self-modify in response to environmental stimuli or user input – core features of DILS. In this way, BlueCity Lab advances biodesign for circularity and provides the infrastructural, methodological and ecological foundations for the next generation of living interfaces. Combining community-based experimentation with scientific rigor and design thinking exemplifies how local innovation ecosystems can meaningfully contribute to a global regenerative future.

Founded in 2011 by Jalila Essaïdi, the *BioArt Laboratories Foundation*[38] is a multidisciplinary, international non-profit organization based in Eindhoven, Noord Brabant. The Foundation serves as a nexus for talent in design, science, business and academia, fostering cutting-edge innovation at the intersection of technology, nature and creativity. Each year, BioArt Laboratories welcomes a diverse cohort of researchers, professionals and students, ensuring a constant influx of new ideas and perspectives. This vibrant exchange of knowledge has positioned the foundation as a leading hub for cross-disciplinary collaboration and innovation, advancing the integration of science and design in groundbreaking ways.

The foundation is located in a repurposed historic military complex nestled in a dense forest in Eindhoven, providing an inspiring environment for experimentation and development. This unique setting encourages serendipity and creativity, facilitating entrepreneurship and the emergence of new biomaterials and start-ups.

BioArt Laboratories provides access to state-of-the-art laboratory facilities, biomaterials and a network of experts, supporting talent in exploring the boundaries between nature, science and art. Through public events such as lectures, exhibitions and workshops, the foundation engages the wider community and fosters dialog on the relevance and acceptance of innovation in biotechnology and the life sciences. By providing these resources and opportunities, BioArt Laboratories plays a crucial role in nurturing a new generation of innovators capable of addressing contemporary challenges through the fusion of biotechnology and the arts.

From the Netherlands to Spain, where the *Genetic Architectures Research Group*[39] at the Universitat Internacional de Catalunya (UIC) in Barcelona represents a key reference in exploring biological–digital integration in design and architecture. Founded in 2000 and coordinated by Alberto T. Estévez, the group conducts interdisciplinary and international research at the intersection of biology, digital technology and architectural design.

At the heart of the group's mission is to merge living systems with computational and parametric design tools to develop novel applications for the built environment that are both technologically advanced and grounded in philosophical and humanistic thought (Estévez, 2020). Their research addresses biomimetics, genetic design, biofabrication and responsive architectural systems, contributing to a growing body of work that reimagines architecture as a living, evolving entity.

A significant milestone in their development was the creation of BioLab, a facility within the UIC School of Architecture that supports hands-on experimentation at the intersection of synthetic biology, digital fabrication and computational design. Through BioLab, the group is pioneering the integration of biological matter – from cellular systems to genetic information – into interactive and adaptive architectural solutions.

The Genetic Architectures Research Group plays a vital role in advancing the field of DILS by contributing architectural visions that integrate living responsiveness, material intelligence and environmental adaptability. Their work exemplifies how architecture can transcend static forms and actively engage with its users and environment, opening up new paradigms that blur the boundaries between the natural and the artificial, the biological and the digital.

The *Basque BioDesign Centre (BDC),*[40] located in Gueñes, near Bilbao, is a pioneering multidisciplinary center dedicated to the digital transformation of future materials through experimentation, innovation and cross-sector collaboration. Designed as a strategic platform for knowledge and production, the BDC operates at the intersection of creativity, biotechnology, digital manufacturing and sustainability, providing infrastructure and expertise to both educational and industrial stakeholders.

Bridging design, science and business, the center integrates biodesign methodologies with new digital tools such as additive manufacturing, parametric design and advanced prototyping. It supports projects that transform biological resources into novel materials, products and systems in line with regenerative principles and circular economy goals.

What distinguishes the BDC is its emphasis on the productive role of creativity as a driver of innovation – not only in the cultural sector but also in technology-intensive industries. The BDC promotes a vision of design as a transformative agent capable of guiding sustainable technological development across disciplines.

In this context, the BDC makes a meaningful contribution to the development of DILS by creating the conditions for biological intelligence, material responsiveness and co-creation with living systems to be embedded in industrial and social practices. Through its programs, facilities and collaborations, the BDC fosters a new generation of materials and interfaces that are alive, adaptive and deeply intertwined with their ecological and technological contexts.

The *Biolab at Fab Lab Barcelona*[41] was initiated by Anastasia Pistofidou,[42] co-founder of Fab Textiles and Fabricademy,[43] as a space to introduce biodesign and biofabrication. Initially developed to support Fabricademy – a global educational program focused on new textile technologies and sustainable design – the Biolab quickly evolved into a pioneering space for biological experimentation and innovation within the broader framework of distributed design. In 2022, the Biolab was restructured and renamed the Beta Biolab, reflecting its expanded mission and integration within Fab Lab Barcelona's activities under the Institute for Advanced Architecture of Catalonia (IAAC). As part of this broader institutional framework, the Beta Biolab plays a key role in redefining how living systems, technology and design converge in distributed, regenerative and human-centered futures.

Today, the Beta Biolab continues to build upon its original focus on textiles while expanding into a multidisciplinary ecosystem that integrates biological systems into fields such as food, urbanism, architecture and product design. Guided by a regenerative and open-source ethos, the lab seeks to democratize access to biotechnological tools and knowledge and to promote a circular approach to design and material development across sectors.

The Biolab is equipped to support experimentation with microorganisms such as bacteria, fungi, algae and small plants using DIY lab equipment, biohacking techniques and digital fabrication tools such as 3D bioprinters. It provides a platform for researchers, artists, designers and citizen scientists to develop living materials, biofabricated structures and biomimetic systems that challenge traditional material paradigms. Applications range from living artworks and edible materials to biodegradable packaging, mycelium-based composites and responsive textiles. In addition to sustainable materials exploration, Biolab is making a significant contribution to the development of HCI and DILS. By exploiting the responsive properties of living organisms – such as the photosensitivity of algae or the chromatic behavior of bacteria – researchers can create interactive biodigital systems that adapt to environmental stimuli or user interaction. These living biosensors open up the possibility of interfaces that evolve, learn and respond in real time, combining biological adaptability with computational precision.

In Portugal, biodesign and biofabricated materials are becoming increasingly important, thanks to the work of pioneering initiatives and researchers. Among the most notable are the *CoCoon* project[44] and the contributions of Américo Mateus[45] and Carla Paoliello,[46] who are actively promoting the integration of biological processes, material innovation and the design process.

CoCoon[47] is a three-year strategic alliance that combines biology, design, education and infrastructure to promote sustainable and future-oriented creative practices. A key objective of the project is to create biolabs – experimental environments where students, educators and researchers can engage in hands-on exploration and prototyping with living materials. These biolabs support green prototyping and aim to replace fossil-based materials with regenerative, biobased alternatives. The project is highly relevant to the emerging field of DILS. By promoting the creation of biolabs, the development of living materials and the integration of biological processes into design education and practice, CoCoon and its associated researchers are directly contributing to developing regenerative, responsive and systemic approaches to material design. These initiatives help to build both the epistemological and infrastructural foundations for DILS in Portugal and beyond, positioning design not only as a creative activity but also as a catalyst for ecological transformation.

CoCoon also supports the integration of biodesign into academic curricula at all levels – from higher education to vocational and lifelong learning – promoting digital literacy and sustainability-oriented skills. The initiative fosters an innovation ecosystem aligned with the green transition and broader European climate goals.

Based in Écublens, Switzerland, *Hackuarium*[48] is a non-profit association committed to democratizing science through public engagement and fostering a culture of open and participatory research. As a community laboratory for open science, Hackuarium welcomes people from all backgrounds – scientists, engineers, designers, artists and curious minds alike – to explore the frontiers of biology, technology and interdisciplinary innovation.

Hackuarium's mission is to make science accessible to everyone by providing an inclusive and collaborative environment where knowledge can be co-created and shared. The association adheres to the *DIYbio Europe Ethical Guidelines*,[49] ensuring that all projects are carried out responsibly, transparently and with a clear societal impact.

Rooted in the principles of open-source science, Hackuarium emphasizes the development of low-cost, reproducible and simple solutions that can be widely adopted and adapted. By lowering the barriers to entry, the organization empowers individuals and communities to participate meaningfully in scientific research and innovation, paving the way for a more equitable and distributed scientific ecosystem.

There are several leading academic institutions and research facilities in the United Kingdom dedicated to biodesign, which integrate living systems into design and architecture to advance DILS and innovative interaction design.

Central Saint Martins offers a pioneering *Masters in Biodesign*,[50] providing students with an experimental learning environment that combines scientific principles with creative design, exploring bioinformed design strategies as a driver for sustainable innovation. The program emphasizes ecological literacy, bioethics, whole systems thinking and practical laboratory work, enabling students to reimagine materials and processes for a sustainable future. The institution is also home to the *Living Systems Lab*,[51] a groundbreaking research group established at Central Saint Martins in 2013. The lab explores the inherent properties of biological systems and how these can inform the creation of regenerative, sustainable and ethically engaged design practices. Rooted in a transdisciplinary approach, the lab brings together artists, designers, bioengineers, environmental scientists and ecologists to rethink the role of living matter in design.

Led by pioneering Professor Carole Collet, the lab functions as both a research center and a creative incubator, where biological processes are studied and collaboratively cultivated as active design agents. From biotextiles and biofabrication to speculative critical design and ecological futures, the lab fosters cross-species collaboration and positions living systems – such as fungi, algae, bacteria and plants – not as passive resources but as co-creators in the design process.

The *Grow Lab*,[52] always at Central Saint Martins, is a Containment Level 1 biology laboratory designed to support students working at the intersection of design, biology and sustainability. It provides space and equipment to grow, cultivate and experiment with living materials – including fungi, algae and bacterial cultures – within a design context. The lab encourages responsible material innovation and speculative exploration, contributing to developing biofabricated and regenerative design practices and providing essential infrastructure for students involved in the MA Biodesign and other projects related to living systems.

The *Bio-ID Lab*, part of the Bio-Integrated Design (Bio-ID) program at UCL's Bartlett School of Architecture in collaboration with the Department of Biochemical Engineering,[53] represents an innovative frontier in biodesign. Unlike conventional approaches that take inspiration from nature, this program positions nature as an active design medium, directly involved in the formation of materials, structures and interactions.

Using an integrated methodology that combines experimental design, biological engineering and digital simulation, the Bio-ID Lab explores complex, multi-scale biologically, materially and socially embedded systems. Its research encompasses living architecture, biofabrication and responsive systems, pushing the boundaries of how biological processes can influence the production, function and interaction of built environments.

A key contributor to the development of DILS, the Bio-ID Lab investigates how biological entities – such as bacteria, algae and engineered tissues – can be incorporated into interactive design, enabling structures that adapt, grow or communicate with their environment and users. This interdisciplinary convergence fosters a new material culture rooted in sustainability, adaptability and cohabitation with living systems, reshaping how we conceive, build and interact with the spaces around us.

At the University of the West of England (UWE) in Bristol, the *Unconventional Computing Laboratory*[54] was founded in 2001 by Andrew Adamatzky as a pioneering center for exploring the future of computing through nature-based systems. The Laboratory investigates the computational capabilities of physical, chemical and biological media, focusing on slime molds, mycelium and other living substrates, to develop non-linear media-based computing prototypes. These systems are not merely theoretical but designed for interactive use, allowing direct engagement between users and the living medium. By exploiting the unique properties of these organisms – such as their growth patterns, environmental responsiveness and adaptive behaviors – the lab creates biohybrid devices that challenge traditional models of computation and interaction.

Adamatzky's work has extended to the development of living wearable technologies, where biological matter is integrated into computational devices capable of sensing and responding to environmental or bodily signals (Adamatzky et al., 2021). His research into "living architecture" further explores the use of mycelial networks as embedded computing systems, envisioning buildings and structures that process information and interact with their surroundings and inhabitants.[55]

These innovations have significant potential for advancing DILS. By demonstrating that living organisms can act as computational agents, the Lab is redefining the relationship between technology, biology and user interaction, offering new paradigms for how humans engage with the built and digital environment symbiotic and regeneratively.

Another leading center for biodesign research in the United Kingdom is the *Hub for Biotechnology in the Built Environment*

(HBBE),[56] jointly run by Newcastle University and Northumbria University. Founded with the ambition to transform how we design, construct and occupy buildings, the HBBE is pioneering the integration of biotechnology into architectural systems to develop responsive, adaptive and biologically regenerative structures.

One of its most innovative research initiatives is *Human–Bacteria Interfaces (HBI)*,[57] which explores the embedding of living microbial systems in architectural elements. These bio-interfaces, composed of bacterial cellulose and microbial consortia, are designed to sense and respond to changes in the environment – emitting light, changing color or releasing odors in response to stimuli. Such responses create multisensory communication channels between humans and microbial life, challenging conventional boundaries between user and environment. HBI is closely linked to the broader theoretical framework of *Ambient Living Intelligence (ALI)*,[58] which reimagines buildings as living ecosystems that actively interact with their occupants. In this paradigm, microbes act not only as sensors but also as expressive agents, capable of communicating information about air quality, humidity or even user behavior via engineered biosignals such as fluorescence or olfaction.

To support this ambitious research, the HBBE is equipped with an ecosystem of facilities, including macro and micro biodesign labs, a multiomics lab and a hybrid workshop space that bridges biological experimentation and materials fabrication. These infrastructures enable a complete workflow – from genomic sequencing and microbial propagation to robotic construction and environmental simulation – fostering the development of truly interactive living systems.

By embedding biological processes directly into the built environment, the HBBE redefines interaction design in architecture. Its work pushes forward the field of DILS by merging synthetic biology, computational tools and human–computer interaction to create buildings that are not only sustainable but also intelligent, sentient and symbiotic with human and microbial life.

Given its originality and pioneering approach to biotech infrastructure, it's impossible not to mention *OpenCell*[59] in this overview. Based in London, OpenCell is a unique model of modular, digitally integrated and accessible laboratory environments housed in repurposed shipping containers. It offers fully serviced BSL2+ laboratories tailored for start-ups in the fields of synthetic biology, diagnostics, foodtech and biomaterials. By combining flexible wet lab spaces with smart Internet of Things (IoT) infrastructure and bespoke lab management tools, OpenCell fosters agile, distributed biodesign. Its commitment to democratizing scientific infrastructure aligns directly with the ethos of DILS, making it a key reference within the evolving landscape of living systems labs.

Located within the Center Rog[60] in Ljubljana, Slovenia, the *Green Lab*[61] is a pioneering interdisciplinary hub in Eastern Europe that brings together scientists, designers, artists and urban gardeners to collaboratively explore the intersection of biotechnology, sustainable design and ecological innovation. A unique space dedicated to community engagement and advanced experimentation, the lab provides facilities and guidance for working with microorganisms, living materials and circular systems.

More than a makerspace, the Green Lab is a living ecosystem of ideas, enabling projects ranging from the development of biobased and biofabricated materials and urban farming prototypes to citizen science initiatives and transdisciplinary education. It fosters a hands-on, inclusive learning environment that redefines traditional approaches to material innovation by encouraging participants to engage directly with natural systems and regenerative practices.

Its mission is closely aligned with DILS principles: to create responsive, sustainable and socially grounded technologies that foster a deeper connection between humans, microbes and the environments we share.

One of the most pioneering and long-standing biolabs in Australia is *SymbioticA*,[62] based at the University of Western Australia (UWA). Founded in 2000 by Oron Catts and IZurr, the lab operates within the School of Human Sciences but maintains strong links with the life sciences, engineering and the arts. SymbioticA is internationally renowned for integrating wet biology practices with artistic, critical and speculative approaches, establishing itself as a pioneer at the intersection of bioart, biodesign and living systems research.

What sets SymbioticA apart is its radically transdisciplinary model. It is a research laboratory, an educational platform, a residency program and a critical forum for cultural debate on the life sciences. Artists, designers, philosophers and scientists come together to work hands-on with living matter, often in projects that challenge established ethical norms and interrogate the cultural consequences of biotechnology and synthetic life.

SymbioticA has played a formative role in shaping the global discourse on living materials and DILS. Encouraging experiential, embodied practices and critical reflections on the manipulation of life pushes the boundaries of how living organisms are integrated into art and design. The Lab's programming – from symposia and exhibitions to technical training and public dialog – cultivates a space for redefining the relationship between human and non-human agencies, with profound implications for regenerative design, biopolitics and interaction ethics.

Also based in Australia but in Sydney, the *Affective Interactions Lab*[63] is a research initiative led by Phillip Gough, program director of the major in Biological Design at the University of Sydney. The lab explores the emotional, sensory and embodied dimensions of interaction at the intersection of HCI, biological systems and design research. It investigates how artifacts, environments and biological interfaces can be designed to engage affectively with human users – responding to touch, emotion, memory or environmental cues.

Rooted in experimental design and critical inquiry, the lab contributes to the advancement of DILS by integrating emotional intelligence and affective responsiveness into biodigital interfaces. Through speculative design, prototyping and cross-disciplinary collaboration, the lab seeks to create systems that function, feel, express and respond – inviting deeper, more meaningful relationships between humans and living technologies.

In Asia, where digital interaction technologies are highly developed, there is growing interest in integrating biological systems into HCI. However, this practice is still emerging and less widespread than in other regions. Nevertheless, pioneering initiatives are laying the groundwork for biodigital convergence through interdisciplinary experimentation.

One such initiative is *(Art)ScienceBLR*,[64] a public laboratory based at the Srishti Institute of Art, Design and Technology in Bengaluru, India. Founded in 2009, the lab enables artists and designers to engage with scientific practices in biodesign, bioelectronics and synthetic biology and explore their artistic, political and social implications. Located at the intersection of art, science and education, (Art)ScienceBLR promotes interactive, critical engagements with living systems, often through speculative artifacts that allow people to interact with plants, microbes or invisible ecosystems. Its commitment to open, public knowledge-making and education has made it a unique space for biodesign in the Global South, particularly in fostering critical discourse around technology and life.

The *FREAK Lab*[65] (Futuristic Research Cluster of Thailand) is another notable example in Thailand. Within this cluster, the biodigital group explores the intersection of biotechnology and digital technology to address pressing global and local challenges. Their work includes developing biodigital systems as interactive tools while interrogating the human body's biological processes as computational interfaces. Projects such as *Hormone Couture* (Pataranutaporn et al., 2017) illustrate the lab's interest in blurring the boundaries between living and non-living matter, raising important ethical questions around bioagency, bodily data and biopolitical futures. Through speculative design and participatory research, the biodigital group situates itself at the edge of bioethics, affective computing and biofabrication, advancing the discourse on DILS in Southeast Asia.

Based in Tokyo, *BioClub*[66] is a vibrant DIY bio-community lab for designers, artists, scientists and citizens exploring the intersection of biology, art and design. Founded in 2015 by visual artists Georg Tremmel and Shiho Fukuhara, the initiative grew from their shared interest in combining bioinformatics, critical design and aesthetic experimentation to create meaningful bioartworks and speculative prototypes. Established in collaboration with the creative co-working platform Loftwork, the lab is located in the heart of Shibuya, Tokyo's cultural and innovation district.

BioClub is an open, inclusive space for hands-on biological research, creative prototyping and transdisciplinary dialog. It offers a rich program of workshops, exhibitions, talks, residencies and collaborative research events that connect local and global communities interested in the cultural, political and practical dimensions of biotechnology.

A key strength of BioClub is its ability to bridge critical inquiry and innovation, supporting projects ranging from speculative biodesign to start-up ventures and public engagement. Its activities support the development of biofabrication techniques, living interfaces and biological sensing systems, contributing directly to the discourse on DILS. By encouraging community-driven exploration of living systems, BioClub plays an essential role in democratizing access to biotechnology and rethinking how life can be ethically and creatively integrated into contemporary design practice.

In Yogyakarta, Indonesia, the *House of Natural Fiber (HONF) Foundation*[67] is a pioneering new media art laboratory founded in 1999 on critical innovation, community engagement and transdisciplinary practice principles. Led by interior designer Venzha Christ, in collaboration with Irene Agrivina (fashion designer), Istasius Sulaiman (graphic designer) and Tommy Surya (VJ and media artist), HONF brings together diverse creative disciplines to explore the social and technological implications of science and art in everyday life. Since its inception, HONF has functioned as a hybrid platform for cultural development, producing a wide range of citizen laboratories, experimental courses, multimedia installations and public exhibitions. The foundation's approach emphasizes education, critical media literacy and the localization of knowledge, often working with marginalized communities to bridge the gap between high technology and grassroots innovation.

A notable example of HONF's bioart practice is the 2011 project *Intelligent Bacteria – Saccharomyces cerevisiae*, created in collaboration with the Microbiology Department of Gadjah Mada University.[68] This project aimed to raise awareness of safe fermentation technologies, particularly in response to the health hazards associated with producing homemade alcoholic beverages. Combining biological research, sensor technology and sound art, the installation featured a network of interconnected bottles of fermenting liquids, each equipped with sensors and microphones that translated microbial activity into an evolving sonic experience. This work exemplifies HONF's method of translating scientific processes into immersive, accessible and participatory systems.

In the context of DILS, HONF represents a critical and culturally grounded model of biodigital experimentation. Their work demonstrates how living organisms, sensory interfaces and community education can come together to produce speculative yet socially impactful artifacts, particularly in non-Western contexts where traditional knowledge and new media can meaningfully intersect.

In China, the field of biodesign and living systems is developing rapidly, supported by government and academic initiatives promoting biotechnology integration into industry and design. The Chinese government's 14th Five-Year Plan for the Development of the Bioeconomy (2021–2025) highlights the importance of biomedicine, organic agriculture, biomanufacturing and biosecurity as strategic sectors for sustainable economic growth (Zhang et al., 2022).

A remarkable example is Tongji University in Shanghai, which actively participates in the Biodesign Challenge and hosts the *Biomimetic Design Lab (BiDL)*,[69] which is dedicated to biomimicry and nature-inspired innovation. The lab was also involved in the publication of *The Rise of Biodesign*, edited by Mary Polites in 2019 (Polites, 2019), the result of three years of research and collaboration with experts and institutions in China focused on bioinspired design. The book introduces the umbrella term "bioinspiration" and clarifies the distinctions between its sub-fields, such as biomimetics and biomimicry. Among the university labs profiled are the Key Laboratory of Bionic Engineering at Jilin University, the Key Laboratory of Bioinspired Smart Interfacial Science and Technology at Beihang University, the Dynamic Technology Lab for Bionic Materials and Structures at Donghua University and the aforementioned BiDL at Tongji University.

At *Xi'an Jiaotong University*, new laboratories for biodesign and biofabrication are being developed in collaboration with the Design Department of the Politecnico di Milano as part of a joint school initiative to promote academic exchange and collaborative research.[70]

In Shenzhen, the Southern University of Science and Technology (SUSTech) has established the School of Design, where scholars established the *Materialities Research Group*.[71] This group explores materials through interdisciplinary and hands-on design practices, integrating ecology, technology and culture.

The biodesign landscape in China is evolving rapidly, and, likely, we will soon see the emergence of more laboratories dedicated to biotechnology and biofabrication, useful for specifically supporting design research and the development of DILS.

Located in the heart of the Los Ríos region of Chile, *LabVA*[72] in Valdivia is a pioneering materials innovation laboratory dedicated to reimagining the future of fashion and textiles through biomaterials, regenerative design and circular economy principles. A hybrid space that combines research, education and industry collaboration, LabVA maintains an independent, self-governing and open-source identity, providing a dynamic platform for the exploration of next-generation territory-based materials and alternative production methods (Weiss Münchmeyer & Besoain Narvaez, 2022).

The lab actively explores biobased and biofabricated solutions such as bacterial cellulose, algae-derived fibers and textiles made from agricultural waste, fostering interdisciplinary collaborations that bridge design, science and manufacturing. With a strong focus on material storytelling and transparency, LabVA helps brands and designers implement low-impact practices that align with global sustainability goals.

At a time when biolabs are gaining momentum in South America and globally, LabVA exemplifies a forward-thinking approach to material development that combines academic research with real-world applications to drive transformative change in the fashion and textile sectors. This initiative undoubtedly provides a solid foundation for future research in the field of DILS, offering valuable insights into the integration of sustainability, technology and design innovation.

Shifting our focus from Chile to Colombia, we encounter another key laboratory in the Latin American biodesign landscape: the *Natural Materials Lab.*[73] Based at the Universidad de los Andes in Bogotá, Colombia, and developed in collaboration with Columbia GSAPP, the Natural Materials Lab is a space for research and experimentation that explores natural materials' architectural and environmental potential, with a particular focus on Latin American contexts. The lab integrates scientific, computational and craft-based approaches to investigate the performative and regenerative properties of materials such as raw earth, plant fibers, mycelium and local biomass. Its mission is to rethink the relationship between the built environment and the landscape by promoting a material-centered design practice that is situated, resilient and culturally responsive.

Through prototyping activities, workshops and international collaborations, the Natural Materials Lab contributes to educating a new generation of designers capable of critically engaging with local resources and transforming them into innovative and sustainable building systems. The work of the lab is highly relevant to the DILS framework, as it explores the substrates of living materials and promotes a culture of materiality rooted in biological organisms.

In general, we can see that a growing number of initiatives across South America are shedding light on the increasing relevance of biofabricated materials and the evolving role of designers in what has traditionally been a science-driven field. Among these, the *Matter of South – Biomaterial Cultures from Latin America*[74] project significantly recognizes the richness, complexity and contemporaneity of biomaterial practices across the region. Sponsored by the Matters of Activity cluster, this initiative celebrated locally situated materials, techniques and knowledge systems, bringing to the fore design and craft practices deeply rooted in the biodiversity and ecological systems of the Global South. These practices now intersect with the frontiers of material science, design research and sustainability. The dialog between Indigenous epistemologies and speculative contemporary design creates fertile ground for new forms of material ecology. Platforms such as Matter of South, especially in conjunction with initiatives such as **the Latin American Biodesign Hub**, provide a coherent and visionary framework for building a locally grounded yet globally connected DILS network.[75]

Also contributing to this vibrant landscape is the *Sistemas Materiales Group*,[76] an interdisciplinary experimental research collective based in Buenos Aires, Argentina. The group works at the intersection of design, science and biology to explore radical strategies for material production. Using materiality as the design process's starting point, the group develops bioinspired and responsive materials, integrating both traditional and advanced technologies – including the direct use of living organisms as agents of material fabrication. Their practice spans applied research and critical speculation, aiming to generate new spatial and constructive logic. Their vision – biological, organic and generative – resonates with the reflections emerging from 4.0[77] and makes a meaningful contribution to the international discourse on the organism's role as co-author in the design process. In 2023, the group organized the *Trazos Biomateriales*[78] exhibition in Buenos Aires, which brought together projects and ideas exploring the integration of biological systems in design. Showcasing unconventional materialities developed from, with and for nature, the exhibition highlighted the potential of living systems as creative partners in shaping new material futures.

The global expansion of biodesign laboratories and research initiatives reflects a profound shift in how we conceptualize design, interaction and materiality. From academic institutions in the United States and Europe to grassroots movements in Latin America, Asia and Australia, a growing network of biolabs is redefining the boundaries between living systems and design practice. Whether rooted in art, science or technology, these labs serve as critical hubs for developing DILS, promoting new paradigms of cohabitation, symbiosis and regenerative material culture.

This survey of global biolabs reveals a vibrant and innovative landscape where universities, research institutions and bottom-up initiatives actively create spaces for interdisciplinary work at the intersection of biology, interactive technologies and design. Given the relative youth of the discipline, access to appropriate infrastructure and the development of effective frameworks for experimentation and collaboration remain key challenges. However, around the world, a thriving community of researchers, designers and entrepreneurs is pushing the boundaries and redefining what is possible in biodesign and DILS.

As this chapter has shown, DILS is not a singular movement but a distributed and evolving ecosystem of practices, infrastructures and epistemologies – one that is as diverse as it is interconnected. The emergence of these biolabs marks the beginning of a new design ontology in which life itself becomes an active partner in the creative process. While this mapping can only provide a snapshot of a rapidly expanding field, it signals the dawn of a new era at the intersection of biology, design and technology.

KEY REFERENCES FOR EXPLORING BIODESIGN AND BIOFABRICATION PRACTICES

For designers, researchers and students interested in exploring biodesign and biofabrication and also considering the creation of Designed Interactive Living Systems, several key platforms and initiatives offer valuable entry points into this evolving field. These resources support not only technical learning and creative inspiration but also access to global practice communities and collaboration opportunities.

Among the major education and research platforms in biodesign and biofabrication, several initiatives stand out for their ability to bridge science, design and hands-on experimentation. From global competitions and academic labs to industry-led innovation platforms, these programs provide accessible entry points for students, researchers and creatives looking to explore the intersection of biology and design. These are the resources that we believe are actively shaping the next generation of biodesigners:

The Biodesign Challenge[79] is one of the most influential global educational initiatives that connects university students with mentors from academia, industry and the life sciences. Each year, student teams worldwide develop speculative yet scientifically sound prototypes that integrate emerging biotechnologies and critical design. Finalists present their work at the Museum of Modern Art (MoMA) in New York, positioning the Challenge as a key entry point into the world of biodesign. The 10th edition of the Challenge will take place in 2025.

Biofabricate[80] is a global platform and annual summit founded by designer Suzanne Lee. It brings together researchers, entrepreneurs, investors and creatives who are committed to developing sustainable biomaterials. The event explores the future of biofabrication and living materials, bridging science, fashion, architecture and innovation to reimagine material systems for a regenerative future.

The *Biodesign Academy*[81] provides intensive, hands-on training for those wishing to develop skills in biodesign, synthetic biology, biofabrication and living materials. The program includes workshops, prototyping sessions, seminars and mentoring opportunities with leading international experts – all within a collaborative, transdisciplinary framework. The newsletter is exciting and full of helpful information.

Grow.bio,[82] a project by Ecovative Design, provides educational tools and DIY kits for growing mycelium-based materials. Aimed at makers, students and designers, the initiative democratizes access to biofabrication, enabling users to grow biodegradable and modular artifacts using fungi and agricultural waste.

Then there are community bio and DIYbio initiatives aiming to democratize access to biotechnology by making laboratory environments, tools and knowledge available to a broader audience beyond institutional settings. These spaces foster experimentation, artistic exploration, education and citizen science, often blending open-source culture with scientific inquiry. Whether you are an artist, designer, researcher or simply curious about working with living systems, these platforms provide inclusive, hands-on environments to explore biodesign and biofabrication in a participatory way.

Founded in 2009, *Genspace*[83] is the world's first community biology lab, based in Brooklyn, New York. It provides a safe, inclusive environment where artists, designers, scientists and curious citizens can learn, create and experiment with biotechnology. Offering workshops, public programs, residencies and access to Biosafety Level 1 laboratory facilities, Genspace promotes accessible life science education and biodesign innovation.

Hackteria[84] is a global open-source platform and community dedicated to bioart, biohacking and do-it-yourself biology. It offers collaborative projects, hands-on tutorials and theoretical resources that bridge biology, technology, art and open science. Founded in 2009, Hackteria supports a decentralized network of labs and practitioners experimenting with living systems.

BioMADE[85] (Bioindustrial Manufacturing and Design Ecosystem) is a manufacturing innovation institute funded by the US Department of Defense. It brings together industry, academia and government to scale up bioindustrial manufacturing and promote sustainable biofabrication. BioMADE supports education, workforce development and commercialization of biobased products.

Regarding projects and critical platforms, they provide curatorial, conceptual and artistic insights into the field of biodesign, challenging dominant narratives and offering speculative, inclusive and culturally grounded perspectives on emerging biotechnologies. These initiatives help to shape the ethical, aesthetic and socio-political dimensions of biodesign. Case studies include the following:

Material Factors[86] is a research and exhibition platform exploring the cultural, political and environmental dimensions of materials in biodesign. It offers a growing archive of projects and critical essays and promotes interdisciplinary exchange between designers, scientists and curators.

A Dutch biodesigner and founder of BlueCity Lab, *Emma van der Leest* built her website[87] to showcase her portfolio of speculative and scientific design experiments, educational resources and public lectures. Emma's work focuses on microbial materials and biofabrication processes.

A design studio founded by Natsai Audrey Chieza that explores the intersection of biotechnology, design and culture, *Faber Futures*[88] works with institutions and industry to re-imagine biomaterial development through an inclusive and systemic lens.

Making Kin[89] promotes feminist and ecological approaches to imagining post-anthropocentric futures. An experimental platform, it combines speculative storytelling, multispecies thinking and biotechnology.

Finally, publications and knowledge hubs provide accessible ways to keep up to date with current research, job opportunities and peer-reviewed evidence. These platforms support career development, critical reflection and the dissemination of biodesign as a transdisciplinary practice.

Biodesign Job[90] is a job board dedicated to careers in biodesign and biofabrication, connecting professionals, researchers and students with opportunities worldwide. It includes listings in academia, industry and speculative design.

BioDesign Research Journal[91] is an open-access, peer-reviewed journal that publishes cutting-edge research in biodesign, synthetic biology, and materials innovation. It serves as a critical knowledge base for scholars and practitioners in the field.

CONCLUSIONS

Biolabs dedicated to experimentation with living systems – including interactive living systems and hybrid biodigital systems – are emerging as pioneering spaces where biotinkering, creative experimentation and transdisciplinary collaboration converge. These environments explore the integration of living organisms and biological processes into design and technology, pushing the boundaries of interaction design, product development and material innovation. By fostering collaboration between designers, scientists and engineers, these labs enable the creation of advanced systems that merge biological intelligence with technological components, redefining the traditional boundaries of design and engineering.

Looking ahead, the future of innovation in these labs may be characterized by the rise of biofoundries: highly specialized, automated manufacturing facilities based on synthetic biology and bioengineering. These platforms have the potential to revolutionize the prototyping and scaling of interactive living systems, enabling new biofabrication techniques and the development of biologically based artifacts and systems, from biosensors, biomaterials and pharmaceuticals to living circuits, biofuels and next-generation interfaces (Hillson et al., 2019; Holowko et al., 2021). These innovations promise to combine the adaptability of biology with the precision of computation, opening up radically new paradigms of responsiveness and functionality.

This survey of biolabs has provided a comprehensive view of a rapidly evolving landscape at the intersection of design, biology and technology. Despite the diversity of contexts and approaches, many of these labs share many defining characteristics:

- A strong commitment to transdisciplinary collaboration, integrating biology, engineering and design to address contemporary societal and environmental challenges
- A focus on interactive living systems that dynamically interact with their environment, often through biologically driven computation or feedback mechanisms
- A commitment to bridging research and application, enabling experimental projects to evolve into deployable, market-ready solutions that can redefine industrial and social practices

Far from being limited to material experimentation, biolabs represent a new frontier in the design of interactive and hybrid systems, where living and technological elements work in synergy. By enabling the development of adaptive, resilient and ecologically integrated systems, these labs are shaping the future of design and innovation and redefining our relationship with matter, technology and the living world.

Note to the Reader

We have made every effort to map the biolabs presented in this chapter, although we are aware that some may have been unintentionally omitted. The biodesign landscape is constantly evolving, with new initiatives and laboratories emerging around the world.

To support ongoing research and improve visibility, we have created an interactive map which will soon be available at www.livinginteractiondesign.com, a website which is currently under development (October 2025).

We plan to keep this map updated and growing. If you know of other relevant labs or initiatives, we invite you to share them with us by contacting our team via the Materials Design for Transition Instagram account (@materialsfortransition_polimi).

Your contributions are essential to help us grow and maintain this open, collaborative resource.

NOTES

1 The Edna W. Lawrence Nature Lab at RISD, School of Design, Rhode Island, US: https://naturelab.risd.edu/
2 MIT Media Lab: The Media Lab is a dynamic interdisciplinary creative sandbox, firmly grounded in academic excellence, where innovation thrives through the collaboration of dozens of research groups, initiatives and centers on hundreds of transformative projects. https://www.media.mit.edu
3 https://wearablebiotech.media.mit.edu/
4 https://tangible.media.mit.edu/
5 https://www.media.mit.edu/groups/fluid-interfaces/overview/
6 Arizona State University's Center for Science and the Imagination brings writers, artists and other creative thinkers into collaboration with scientists, engineers and technologists to reignite humanity's grand ambitions for innovation and discovery. The center serves as a network hub for audacious moonshot ideas and a cultural engine for thoughtful optimism. They provide a space for productive collaboration between the humanities and the sciences, bring human narratives to scientific questions and explore the full social implications of cutting-edge research. https://science.asu.edu/center-science-and-imagination

7 Living Matter Lab, The ATLAS Institute, College of Engineering and Applied Science, University of Colorado Boulder. Available at: https://www.colorado.edu/atlas/living-matter-lab
8 Mirela Alistar, Assistant Professor at the ATLAS Institute and the Department of Computer Science at the University of Colorado Boulder, leads the Living Matter Lab and teaches Introduction to Biodesign and Physical Interaction with Living Matter. https://www.personallab.org/
9 BioInteractive Design (BInD) Lab, Northeastern University, Interdisciplinary Science & Engineering Complex (ISEC). Available at: https://biointeractive.lab.northeastern.edu/
10 Designing pARticipatory futurEs (DARE) Lab, Department of Information Systems, University of Maryland, Baltimore County. Available at: https://www.participatoryfutures.com/
11 More information available at: https://umbc.edu/stories/foad-hamidi-launches-new-projects-to-expand-technology-rich-learning-opportunities-for-youth-in-baltimore/
12 More information available at: https://informalscience.org/project/introducing-synthetic-biology-using-co-designed-culturally-responsive-biomaker-activities-for-family-engagement-in-underserved-communities/
13 INCUBATOR Art Lab, University of Windsor. A hybrid BSL-2 laboratory and creative space founded by Jennifer Willet, supporting bioart research, education and public engagement at the intersection of art, science and ecology. Available at: https://incubatorartlab.com/about/
14 BioLab, Burg Giebichenstein University of Art and Design Halle. Available at: https://www.burg-halle.de/en/university/facilities/burglabs/biolab
15 Mareike Gast, "*Products Will Grow in Labs* – The BURGLabs," published on 24 November 2022. Available at: https://ndion.de/en/mareike-gast-products-will-grow-in-labs-the-burglabs
16 Bio Design Lab, Hochschule für Gestaltung Karlsruhe (HfG Karlsruhe). Available at: https://biodesignlab.hfg-karlsruhe.de/
17 *Matters of Activity. Image Space Material*, Cluster of Excellence at Humboldt University in Berlin. Available at: https://www.matters-of-activity.de
18 __matter Festival 2025, organized by the Matters of Activity Cluster of Excellence. A multi-venue event in Berlin exploring material agency through exhibitions, workshops, and debates. More information available at: https://www.matters-of-activity.de/en/matter-festival
19 Laboratorium, based at the Media Arts Studio of KASK / School of Arts Ghent. Available at: https://www.laboratorium.bio
20 ETHO Online Session: New Material Practice in Biolabs, hosted by ELIA (March 2025). More information at: https://elia-artschools.org/page/ETHOOnlineSessionNewMaterialPracticeinBiolabs
21 The Centre of Design Research for Regenerative Material Ecologies (DREAM), Faculty of Industrial Design Engineering, TU Delft. Available at: https://www.regenerativematerialecologies.com

22 TU Delft. (2022). Opening of TU Delft's state-of-the-art Biodesign Lab. https://www.tudelft.nl/en/2022/io/december/opening-of-tu-delfts-state-of-the-art-biodesign-lab
23 More information available at: https://www.nextskins.eu/
24 Material Experience Lab, Faculty of Industrial Design Engineering, TU Delft and Design Department, Politecnico di Milano. Available at: https://materialsexperiencelab.com/
25 The development of the Biolab at the Department of Design of the Politecnico di Milano has been made possible by the participation in the MIcS – Made in Italy Circular and Sustainable project, funded by the Italian National Recovery and Resilience Plan (PNRR). In particular, the initiative is supported by Spoke 5 – 03: FasT4C – Fashion-Tech Design for Circularity (2023–2025), coordinated by Prof. Paola Bertola and Daria Casciani, which promotes research infrastructure for sustainable innovation in fashion and design. Available at: https://www.mics.tech/projects/5-3-fast4c-fashion-tech-design-for-circularity
26 The Materials Design for Transition research group at the Politecnico di Milano has an active presence on Instagram under the handle @materialsfortransition_polimi. The account showcases projects and initiatives at the intersection of biodesign, sustainability and materials innovation.
27 Hybrid Design Lab, originally established at the Second University of Naples and now based at the University of Naples Federico II, under the coordination of the founder, Professor Carla Langella. Available at: https://www.hybriddesignlab.org
28 BIOlogic is the technology laboratory of Knowledge for Business https://www.kforbusiness.it/biologic/
29 Sustainable Design Laboratory, University of Florence. https://www.dida.unifi.it/vp-479-sustainable-design-laboratory.html, together with the research unit in Biodesign: https://www.dida.unifi.it/vp-941-biodesign.html#
30 Material Aesthetics Lab, Eindhoven University of Technology. Facilities include the Bio Lab (ML-I), additive fabrication equipment and interdisciplinary collaboration spaces. Available at: https://materialaesthetics.com
31 The Bio Lab (ML-I), a part of the Material Aesthetics Lab, is a dynamic environment designed for interdisciplinary biomaterial explorations. It specializes in aseptic conditions for preparing, growing and post-processing material samples. The lab advances possibilities of additive manufacturing through two multifunctional and hackable bio printers. https://research.tue.nl/en/equipments/bio-lab
32 Material Incubator (MI) is an interdisciplinary creative research lab that explores living and creating in collaboration with living (micro) organisms through artistic, scientific and design-based approaches. See: https://materialincubator.com/materialincubator/
33 Waag is a foundation supported during the 2021–2024 period by the Cultural Basic Infrastructure (BIS) of both the Dutch Ministry of Education, Culture and Science and the City of Amsterdam. Waag has

officially been designated as a Futurelab by the Ministry. Available at: https://waag.org/en/

34 Open Wetlab, Waag Futurelab. A public biolab exploring biodesign, bioart and participatory biology at the intersection of science, society and art. Available at: https://waag.org/en/lab/open-wetlab/

35 BioHack Academy, Waag Futurelab. An international educational program that teaches participants how to build biolabs, conduct biological experiments and critically engage with synthetic biology using open-source tools. Available at: https://waag.org/en/project/bio hack-academy/

36 Hybrid Lab Network. A European network connecting interdisciplinary laboratories at the intersection of art, science and society. Available at: https://hybrid.i3s.up.pt/

37 BlueCity Lab, Rotterdam. A circularly built biolab supporting biodesign, entrepreneurship and experimentation with living materials for a regenerative and circular economy. Available at: https://www.bluecity .nl/en/blue-city-lab

38 BioArt Laboratories Foundation, Eindhoven. A multidisciplinary non-profit organization founded by Jalila Essaïdi, fostering innovation at the intersection of art, science and nature. Available at: https://bioartlab. com/about/organisation

39 Genetic Architectures Research Group, Universitat Internacional de Catalunya (UIC), Barcelona. Founded in 2000 and coordinated by Alberto T. Estévez, the group explores the convergence of biology, digital technologies and architecture. Available at: https://geneticarch itectures.weebly.com; https://www.uic.es/en/research/research/resea rch-groups/engineering-and-architecture/grc-genetic-architectures-re search

40 Basque BioDesign Center (BDC), Gueñes, Bilbao. A multidisciplinary innovation hub focused on biodesign, digital manufacturing and sustainability through creativity and experimentation. Available at: https://www.basquedesigncenter.com/

41 Beta Biolab at Fab Lab Barcelona. An open biolab dedicated to exploring biodesign, biofabrication and the integration of living systems into diverse design disciplines. Available at: https://fablabbcn.org/blog /lab-life/inside-the-biolab

42 Anastasia Pistofidou is a researcher and practitioner in digital fabrication and biodesign, part of the team at Fab Lab Barcelona. See: https://fablabbcn.org/team/anastasia-pistofidou

43 The Textile Academy is an international educational program focusing on textiles, soft fabrication and biobased materials, bridging craft, digital fabrication and sustainability. Available at: https://textile-ac ademy.org/

44 CoCoon (2023–2025) is a Forward-Looking Project funded under the EU Erasmus+ Programme – Partnerships for Innovation. The project aims to foster innovation, creativity, and entrepreneurial skills through the development of green and biodesign competencies. Its focus lies in promoting "learning by doing" approaches to enhance green prototyping and the effective use of digital technologies. See: www. cocoon.bio

45 Américo Mateus is a researcher and innovation strategist involved in participatory design, sustainable futures and co-creation methodologies. He is affiliated with Universidade Lusófona and the URBiNAT project. See: https://urbinat.eu/people/americo-mateus/

46 Carla Paoliello is an Invited Assistant Professor at the Fine Arts Faculty of Lisbon University and at ISMAT in Portimão. Her research focuses on collaboration and co-design, craft and traditional practices and eco-consciousness and care, as well as identities and territories. She is currently investigating knowledge exchange and experimenting with living systems and biomaterials. See: https://orcid.org/0000-0003-0186-0507 and @carlapaoliello.

47 CoCoon – Co-Creating Greener Futures: Developing and Transferring Innovative Bio-Design Modules for Education to Accelerate the Green Transition. Erasmus+ Programme (ERASMUS-EDU-2022-PI-FORWARD-LOT1), Type of action: ERASMUS Lump Sum Grants, Grant Agreement No. 101087204. Project website: https://www.cocoon.bio. EU Project page: https://ec.europa.eu/info/funding-tenders/opportunities/portal/screen/opportunities/projects-details/43353764/101087204

48 Hackuarium, Écublens, Switzerland. A non-profit community laboratory dedicated to open science, public engagement and democratizing access to biotechnology and interdisciplinary innovation. Available at: https://www.hackuarium.ch

49 DIYbio Europe Code of Ethics. A set of ethical guidelines developed during the European Congress on DIY Biology, promoting responsible, peaceful and accountable use of biotechnology with respect for all living systems. Available at: https://diybio.org/codes/draft-diybio-code-of-ethics-from-european-congress

50 Postgraduate MABiodesign, UAL: https://www.arts.ac.uk/subjects/textiles-and-materials/postgraduate/ma-biodesign-csm

51 The Design and Living Systems Lab is a pioneering research laboratory that explores the interface of biological sciences and design to challenge established paradigms and envision new sustainable materials and forms of production for the future. http://www.designandlivingsystems.com/about/

52 Grow Lab, Central Saint Martins, University of the Arts London. A dedicated facility supporting experimentation with living materials, biofabrication and regenerative design practices within a creative academic context. Available at: https://www.arts.ac.uk/colleges/central-saint-martins/student-life-at-csm/facilities/grow-lab

53 Bio-Integrated Design (Bio-ID) MSc/MRes, UCL Bartlett School of Architecture in collaboration with the Department of Biochemical Engineering. An interdisciplinary program that integrates biological systems into architectural and material design through experimental, scientific and digital methodologies. Available at: https://www.ucl.ac.uk/prospective-students/graduate/taught-degrees/bio-integrated-design

54 Unconventional Computing Laboratory, University of the West of England, Bristol. A research center dedicated to exploring

next-generation computational systems using living, chemical and physical media. Available at: https://uncomp.uwe.ac.uk/

55 FUNGAR – Fungal Architectures. A European cross-disciplinary research project exploring the use of fungal mycelium as a living structural and computational material for architecture. The project investigates how mycelium can enable self-growing, sensing and adaptive built environments. Available at: https://www.fungar.eu

56 Hub for Biotechnology in the Built Environment (HBBE), Newcastle University and Northumbria University. A research initiative aiming to develop biointegrated buildings by merging biotechnology with architecture to create responsive, living environments. Available at: http://bbe.ac.uk/

57 Human–Bacteria Interfaces (HBI), part of the Hub for Biotechnology in the Built Environment (HBBE). A research project exploring the integration of microbial systems into building materials to create responsive, sensory interfaces within architectural environments. Available at: http://bbe.ac.uk/human-bacteria-interfaces/

58 "On Prototyping an Ambient Living Intelligence", *Driving the Human Journal*. An article exploring the concept of ambient living intelligence (ALI), where microbial systems are integrated into built environments to enable responsive, multi-sensory interactions between architecture, humans, and living matter. Available at: https://drivingthehuman.com/journal/on-prototyping-an-ambient-living-intelligence

59 OpenCell, London. A modular biotech innovation hub offering accessible BSL2+ lab space and digital infrastructure for startups. It promotes distributed and agile biodesign. Available at: https://www.opencell.bio

60 Center Rog, Ljubljana, Slovenia. A cultural and creative hub dedicated to design, innovation and sustainable development, offering spaces for makers, researchers and the public to collaborate across disciplines. Available at: https://center-rog.si/en/

61 The Green Lab, located at Center Rog in Ljubljana, Slovenia, serves as a platform for collaborative research and education at the intersection of science, design and nature. Available at: https://center-rog.si/en/labs/green-lab/

62 SymbioticA, University of Western Australia. An internationally recognized artistic research laboratory. It serves as a platform for interdisciplinary experimentation, public engagement and ethical reflection on biotechnology. Available at: https://static.weboffice.uwa.edu.au/archive/www.symbiotica.uwa.edu.au/home.html

63 Affective Interactions Lab, University of Sydney. A research group led by Phillip Gough, focusing on the emotional and sensory dimensions of human interaction with biological systems, artifacts, and technologies – at the intersection of design and HCI. Available at: https://design.sydney.edu.au/research-2/affectiveinteractions/

64 (Art)ScienceBLR, Srishti Institute of Art, Design and Technology, Bengaluru, India. A public laboratory exploring the intersection of art, science and pedagogy through critical engagements with biodesign, bioelectronics and synthetic biology. Available at: https://artscienceblr.org/index.php/about/

65 FREAK Lab – Futuristic Research Cluster of Thailand. A transdisciplinary research platform exploring the convergence of biotechnology, digital technology and speculative design, with initiatives such as the biodigital group addressing emerging biodigital futures. Available at: https://freaklab.org/

66 BioClub Tokyo. A DIY bio community lab located in Shibuya, it brings together artists, designers and scientists to explore the intersections of biotechnology, art and design through workshops, events and collaborative projects. Available at: https://bioclub.tokyo/

67 House of Natural Fiber (HONF), Yogyakarta, Indonesia. A new media art laboratory founded in 1999, dedicated to critical innovation, citizen science and the intersection of art, technology and community engagement. Available at: https://archive.transmediale.de/content/honf-the-house-of-natural-fiber-0

68 *Intelligent Bacteria – Saccharomyces cerevisiae* is a bioart project by the HONF (House of Natural Fiber) laboratory that explores fermentation as a cultural and technological practice, integrating art, microbiology and sensor-based interaction. Documented by Transmediale. Available at: https://archive.transmediale.de/content/intelligent-bacteria-saccharomyces-cerevisiae

69 Biomimetic Design Lab (BiDL), Tongji University. Available at: https://bidl.tongji.edu.cn/

70 Joint School of Design and Innovation Centre. Available at: https://www.polimi.it/il-politecnico/progetti-di-ateneo/joint-school-of-design-and-innovation-centre

71 Materialities Research Group, School of Design, Southern University of Science and Technology (SUSTech). Faculty page: https://designschool.sustech.edu.cn/about/team/faculty/424.html, and https://www.enzamigliore.com/

72 LabVA – Laboratorio de Biomateriales de Valdivia. Available at: http://www.labva.org/ and https://www.rawassembly.com/labva

73 The Natural Materials Lab at Columbia GSAPP explores the architectural, environmental and cultural potential of natural materials through interdisciplinary research and experimentation. The lab investigates materials such as soil, fiber and biobased composites to promote sustainable and locally rooted building practices. See: https://www.arch.columbia.edu/research/labs/17-natural-materials-lab

74 Matters of Activity. (2024). Matter of South – Biomaterial Cultures from Latin America. Available at: https://www.matters-of-activity.de/en/activities/14014/matter-of-south-biomaterial-cultures-from-latin-america

75 The Biodesign Regional Hub – Latin America is a digital space connecting communities of artists, designers and scientists engaged in biodesign practices that work with life and its circular and sustainable properties. Launched jointly by Universidad de los Andes and the Biodesign Challenge (BDC), the hub serves as a platform for practitioners, research institutions and students to exchange resources, ask questions and discuss pedagogical strategies in the fields of biodesign, biotech and bioart. See: https://www.linkedin

.com/groups/12499403 and https://www.biodesignchallenge.org/pressblog/2021/april-15-latam-hub

76 The Sistemas Materiales Group is an interdisciplinary experimental research collective based in Buenos Aires, Argentina. It explores the intersections of design, science and biology to develop bioinspired and responsive materials, integrating both high-tech and low-tech processes, and even using living organisms as material producers and spatial constructors. See: Sistemas Materiales Group. Material Systems. Available at: http://heidijalkh.com

77 The Global Community Bio Summit is a space for the global community of DIY biologists, community biologists, biohackers, biomakers and members of independent and community labs to convene, plan, build community and continue the evolution of the movement. See: Global Community Bio Summit. About the Summit. Available at: https://www.biosummit.org/about

78 The Trazos Biomateriales exhibition, available at: https://biosummit.live/bio-creation-station-3/trazos-biomateriales. More information on the catalogue: Pozzetti, G., & Jalkh, H. (Eds.). (2023). *Trazos: Edición biomateriales. Sistemas Materiales*. ISBN 978-631-00-1112-7

79 Biodesign Challenge. A global competition and education platform connecting students with experts in design and the life sciences to envision the future of biotechnological innovation. Available at: https://www.biodesignchallenge.org

80 Biofabricate. A global summit and knowledge-sharing platform founded by Suzanne Lee to drive sustainable innovation in biomaterials and biofabrication across design, fashion, architecture and science. Available at: https://www.biofabricate.co

81 Biodesign Academy. An intensive educational program offering practical training in biodesign, synthetic biology and biofabrication, aimed at fostering interdisciplinary collaboration and innovation. Available at: https://www.biodesign.academy

82 Grow.bio. An educational platform providing DIY kits and tools for cultivating mycelium-based materials, empowering designers, students and makers to explore sustainable biofabrication. Available at: https://grow.bio

83 Genspace. Community biology lab and education space in NYC. Available at: https://www.genspace.org

84 Hackteria. Open-source biohacking and bioart network. Available at: https://www.hackteria.org

85 BioMADE. Bioindustrial innovation institute in the United States. Available at: https://www.biomade.org

86 Material Factors. A curatorial platform for critical engagement with materials and biodesign. Available at: https://www.materialfactors.org

87 Emma van der Leest. Designer and educator exploring microbial materials. https://www.emmavanderleest.com

88 Faber Futures. Studio exploring biodesign as a tool for systemic cultural change. Available at: https://www.faberfutures.com

89 Making Kin. Platform for multispecies storytelling and speculative biotechnologies. https://makingkin.org

90 Biodesign Jobs. Job platform for careers in biodesign and biofabrication. Available at: https://www.biodesignjobs.com
91 BioDesign Research. Peer-reviewed journal on biodesign innovations. Available at: https://spj.sciencemag.org/journals/bdr

REFERENCES

Adamatzky, A., Nikolaidou, A., Gandia, A., & Chiolerio, A. (2021). Living wearables from slime mould and fungi. *Revue LINKs Series Special Issue*, *1*, 93–98.

Bell, F., Alistar, M., & Buechley, L. (2024a). Integrating biomaterials and interactive technologies: The growth of biodesign within human-computer interaction. In *Research directions: Biotechnology design*. Cambridge University Press. https://doi.org/10.33774/coe-2024-rps9s

Bell, F., Ramsahoye, M., Coffie, J., Tung, J., & Alistar, M. (2023). MMe: Exploring the human microbiome as an intimate material for living interfaces. In *Proceedings of the 2023 ACM Designing Interactive Systems conference (DIS '23)* (pp. 2019–2033). Association for Computing Machinery. https://doi.org/10.1145/3563657.3596133

Bell, F., Wu, S., Campo Woytuk, N., Lazaro Vasquez, E. S., Alistar, M., & Buechley, L. (2024b). Making biomaterials for sustainable, tangible interfaces. In *Proceedings of the eighteenth international conference on Tangible, Embedded, and Embodied Interaction (TEI '24)* (Article 55, pp. 1–5). Association for Computing Machinery. https://doi.org/10.1145/3623509.3634740

Chappell, C. R., Perez, R., & Takara, C. O. (2023). Growing biodesign ecosystems: Community exchange spaces advance biotechnology innovation. *Research Directions: Biotechnology Design*, *1*, 1–8. https://doi.org/10.1017/btd.2023.8

Crawford, A. (2023). *Designer's guide to lab practice*. Routledge. https://doi.org/10.4324/9781003363774

Dade-Robertson, M., & Zhang, M. (2024). Theory and design in the biotechnical age: A schematic understanding of bio design and synthetic biology practice. *The Design Journal*, *27*(5), 800–822. https://doi.org/10.1080/14606925.2024.2381914

de Lange, O., Dunn, K., & Peek, N. (2022). "Short on time and big on ideas": Perspectives from lab members on DIYBio work in community biolabs. In *Proceedings of the 2022 ACM designing interactive systems conference* (pp. 1358–1376). ACM. https://doi.org/10.1145/3532106.3533521

Estévez, A. T. (2020). The fifth element: biodigital & genetics. In A. Vasel-Be-Hagh & D. S. K. Ting (Eds.), Environmental management of air, water, agriculture, and energy (pp. 195–212). CRC Press. https://doi.org/10.1201/9780429196607

Filippi, M., Mekkattu, M., & Katzschmann, R. K. (2024). Sustainable biofabrication: From bioprinting to AI-driven predictive methods. *Trends in Biotechnology, 43*(2), 290–303. https://doi.org/10.1016/j.tibtech.2024.02.005

Forman, J., Pataranutaporn, P., Gough, P., Kim, R., Bell, F., Ofer, N., Lu, J., Vujic, A., Bai, M., Maes, P., Ishii, H., & Sra, M. (2023). Living bits and radical aminos: A workshop on bio-digital interfaces for human-computer interaction. In *Extended abstracts of the 2023 CHI conference on human factors in computing systems* (pp. 1–7). ACM. https://doi.org/10.1145/3544549.3573813

Gabrielian, A. (2024). *Symbiotic futures: Ethical biodesign education in the context of the anthropocene*. Cambridge Open Engage. Retrieved March, 2025, from https://www.cambridge.org/engage/coe/article-details/660cd38f19d524f6bc4f3ac6?show=item

Giaccardi, E., & Redström, J. (2020). Technology and more-than-human design. *Design Issues, 36*(4), 33–44. https://doi.org/10.1162/desi_a_00601

Ginsberg, A. D., Calvert, J., Schyfter, P., Elfick, A., & Endy, D. (2017). *Synthetic aesthetics: Investigating synthetic biology's designs on nature.* MIT Press. https://doi.org/10.1017/S0030605314001045

Ginsberg, A. D., & Chieza, N. (2018). Editorial: Other Biological Futures. *Journal of Design and Science*. https://doi.org/10.21428/566868b5

Groutars, E. G., Kim, R., & Karana, E. (2024). Designing living artefacts for multispecies interactions: An ecological approach. *International Journal of Design, 18*(2), 59–78. https://doi.org/10.57698/v18i2.04

Hamidi, F., & Baljko, M. (2014). Rafigh: A living media interface for speech intervention. In *Proceedings of the 32nd annual ACM conference on human factors in computing systems (CHI '14)* (pp. 1817–1820). ACM. https://doi.org/10.1145/2556288.2557392

Hillson, N., Caddick, M., Cai, Y., Carrasco, J. A., Chang, M. W., Curach, N. C., Bell, D. J., Le Feuvre, R., Friedman, D. C., Fu, X., Gold, N. D., Herrgård, M. J., & Holowko, M. B. (2019). Building a global alliance of biofoundries. *Nature Communications, 10*(1), 2040. https://doi.org/10.1038/s41467-019-10079-2

Holowko, M. B., Frow, E. K., Reid, J. C., Rourke, M., & Vickers, C. E. (2021). Building a biofoundry. *Synthetic Biology, 6*(1), ysaa026. https://doi.org/10.1093/synbio/ysaa026

Ihls, J., & Pollini, B. (2024). *From observation to operation: How lab spaces influence the biodesign practice*. Cambridge Open Engage. https://doi.org/10.33774/coe-2024-rh52p Retrieved March, 2025, from https://www.cambridge.org/engage/coe/article-details/65fa8f3be9ebbb4db9162c36?show=item.

Ihls, J., & Pollini, B. (2025). *From observation to operation: The role of lab spaces in Biodesign practice - Appendix*. Cambridge Open Engage. https://doi.org/10.33774/coe-2025-ww7b0 Retrieved March, 2025, from https://www.cambridge.org/engage/coe/article-details/677d9f796dde43c908a849c5?show=item

Karana, E., McQuillan, H., Rognoli, V., & Giaccardi, E. (2023). Living artefacts for regenerative ecologies. *Research Directions: Biotechnology Design*, *1*, e16. https://doi.org/10.1017/btd.2023.16

Kim, J., Kim, R., Martins, J., & Karana, E. (2024, June 23–28). Becoming microbes: An approach to cultivating microbial sensibilities in biodesign. In C. Gray, E. Ciliotta Chehade, P. Hekkert, L. Forlano, P. Ciuccarelli, & P. Lloyd (Eds.), *Proceedings of DRS2024*. https://doi.org/10.21606/drs.2024.950

Kim, R., Risseeuw, C., Groutars, E. G., & Karana, E. (2023). Surfacing livingness in microbial displays: A design taxonomy for HCI. In *Proceedings of the 2023 CHI* conference *on* human factors *in* computing systems (pp. 1–21). ACM. https://doi.org/10.1145/3544548.3581417

Langella, C., & Fiume, M. (2023). Biodesign: The project that instructs nature. *Cuadernos del Centro de Estudios de Diseño y Comunicación*, *190*, 17–28. https://doi.org/10.18682/cdc.vi190.9530

Langella, C. (2024). Bacteriascape: Synergistic collaborations between design and bacteria. In C. Gambardella (Ed.), *For nature/with nature: New sustainable design scenarios* (Vol. 38, pp. pp 423–439). Springer Series in Design and Innovation. Springer. https://doi.org/10.1007/978-3-031-53122-4_27

Latour, B. (1983). Give me a laboratory and I will move the world. In K. Knorr-Cetina & M. Mulkay (Eds.), *Science observed: Perspectives on the social study of science* (pp. 141–170). Sage. Retrieved March, 2025, from http://www.bruno-latour.fr/sites/default/files/12-GIVE-ME-A-LAB-GB.pdf

Merritt, T., Hamidi, F., Alistar, M., & DeMenezes, M. (2020). Living media interfaces: A multi-perspective analysis of biological materials for interaction. *Digital Creativity*, *31*(1), 1–21. https://doi.org/10.1080/14626268.2019.1707231

Oxman, N. (2016). Age of entanglement. *Journal of Design and Science*, *1*(January). https://doi.org/10.21428/7e0583ad.

Parisi, S., Rognoli, V., & Sonneveld, M. (2017). Material tinkering: An inspirational approach for experiential learning and envisioning in product design education. *The Design Journal*, *20*(sup1), S1167–S1184. https://doi.org/10.1080/14606925.2017.1353059

Pataranutaporn, P., Ngamkajornwiwat, P., Unprasert, T., Umnajsasithorn, W., Luksanayeam, S., Loha-Unchit, S., ... Jain, S. (2017). Hormone couture: Biopolitics, aesthetics, and technology. In *Proceedings of the 2017 ACM International Symposium on Wearable Computers (ISWC)*. ACM. https://doi.org/10.1145/3123021.3123073.

Pataranutaporn, P., Ingalls, T., & Finn, E. (2018). Biological HCI: Towards integrative interfaces between people, computer, and biological materials. In *Proceedings of the 2018 CHI conference on human factors in computing systems (CHI EA '18)* (Paper LBW579, 1–6). Association for Computing Machinery. https://doi.org/10.1145/3170427.3188662.

Polites, M. (Ed.). (2019). *The rise of biodesign*. MAPS Design Studio.

Pollini, B. (2024). From biodesigners to designers in lab: Testing the nuances of an emerging profession through autoethnography. *Research Directions: Biotechnology Design*, *1*, e11. https://doi.org/10.1017/btd.2024.11.

Pozzetti, G., & Jalkh, H. (Eds.). (2023). *Trazos: Edición biomateriales*. Sistemas Materiales.

Raicu, I., & Green, B. (2016). *The ethics of biodesign*. Markkula Center for Applied Ethics, Santa Clara University. Retrieved March, 2025, from https://www.scu.edu/ethics/focus-areas/technology-ethics/resources/ethics-of-biodesign/

Risseeuw, C., Mcquillan, H., Martins, J., & Karana, E. (2024). (Re) activate, (re)direct, (re)arrange: Exploring the design space of direct interactions with Flavobacteria. In *Proceedings of the 2024 CHI conference on human factors in computing systems* (Article 705, pp. 1–18). Association for Computing Machinery. https://doi.org/10.1145/3613904.3642262

Rognoli, V., Pollini, B., & Alessandrini, L. (2021). Design materials for the transition towards post-Anthropocene. In B. Camocini & F. Vergani (Eds.), *From human-centred to more-than-human design: Exploring the transition* (pp. 101–130). Franco Angeli International.

Sawa, M. (2016). The laboratory life of a designer at the intersection with algal biotechnology. *Architectural Research Quarterly*, *20*(1), 65–72. https://doi.org/10.1017/S1359135516000191

Willet, J. (2021, November 8–11). BIOLOVE: Designing hybrid laboratories for creative research at INCUBATOR Art Lab. In *Arts imagining communities to come*, Guayaquil, Ecuador, Cumulus Association.

Weiss Münchmeyer, A. J., & Besoain Narvaez, M. J. (2022). Territory-based biomaterials: Methodology to create a situated biomaterial palette. *Base Diseño E Innovación*, *7*(7), 7–25. https://doi.org/10.52611/bdi.num7.2022.797

Wilson, S. (2023). Bridging science, design, and sustainability with biofabrication. *BioArtelier*. Retrieved March 30, 2025, from https://www.bioartelier.org/post/bridging-science-design-and-sustainability-with-biofabrication

You, W., Valkjärvi, M., & Ofosu, G. (2021). What it takes to make it: Profile and characteristics of DIY biolaboratory founders. *Technology Analysis & Strategic Management*, *33*(10), 1198–1212. https://doi.org/10.1080/09537325.2021.1937978

Zhang, X., Zhao, C., Shao, M. W., Chen, Y. L., Liu, P., & Chen, G. Q. (2022). The roadmap of bioeconomy in China. *Engineering Biology*, *6*(4), 71–81. https://doi.org/10.1049/enb2.12026

Zhou, J., Barati, B., Giaccardi, E., & Karana, E. (2022). Habitabilities of living artefacts: A taxonomy of digital tools for biodesign. *International Journal of Design*, *16*(2), 57–73. https://doi.org/10.57698/v16i2.05

Zhou, J., Doubrovski, Z., Giaccardi, E., & Karana, E. (2024, May 11–16). Living with Cyanobacteria: Exploring materiality in caring for microbes in everyday life. In *Proceedings of the CHI conference on human factors in computing systems (CHI '24)*. ACM. https://doi.org/10.1145/3613904.3642039

CHAPTER 6

DESIGNED INTERACTIVE LIVING SYSTEMS (DILS)

TOWARD A POST-ANTHROPOCENTRIC VISION

DOI: 10.4324/9781003366799-7

A PARADIGM IN TRANSITION

The field of design is undergoing a profound epistemological and ethical shift prompted by environmental crises, technological convergence and growing awareness of interspecies relationships (Escobar, 2018; Haraway, 2016). At the center of this transformation is a challenge to the anthropocentric foundations that have guided design thinking for centuries. As the Anthropocene era reveals the ecological consequences of human-centered development, such as climate disruption, extinctions and toxic waste (Odenthal & Suner, 2017; Morton, 2013), designers must rethink not only what and how they design but also for whom (Tsing, 2015; Tironi et al., 2023; Sánchez Ruano, 2025).

Designed Interactive Living Systems (DILS) embody this paradigm shift. These systems integrate living organisms at the intersection of design and technology. Life becomes a co-performer, collaborator and, sometimes, user. In this context, the dominant human-centered design frameworks, which have long been celebrated for improving usability and user experience, show their limitations. But what happens when the "user" is non-human? What happens when the artifact has its own metabolism, agency and temporal rhythm?

The research on which this book is based investigated how DILS can facilitate the shift from anthropocentric paradigms to multispecies and more-than-human design logic. Drawing on posthuman philosophy (Braidotti, 2013), ecological thinking and the emerging discourse of bioHCI (biological human–computer interaction), the authors argue that DILS are ethical and ontological provocations and design innovations. They encourage designers to move away from control, optimization and functionalism and toward care and cohabitation (Puig de la Bellacasa, 2017; Karana et al., 2020). Central to this vision is the reconfiguration of interactivity itself (Giaccardi & Redström, 2020). Rather than being seen as a loop between user and technology, interactivity in DILS becomes a dynamic negotiation among heterogeneous agents, each with its own intelligence, vulnerability and vitality forms.

This reconceptualization builds on the insights of HCI and UX rather than discarding them. To realize DILS's full potential, we must adopt new methodologies that acknowledge the agency of biological systems, the fragility of interspecies relations and the necessity of designing with livingness, not just for it. By doing so, DILS could help create a more sustainable and symbiotic technological future where design is based on reciprocity and symbiosis.

FROM ANTHROPOCENTRISM TO ECOLOGICAL THINKING

Anthropocentric Design in DILS

Winograd and Flores (1986) argue that early computational systems were modeled on rationalist cognition paradigms, leading to design strategies rooted in centralized control and goal-oriented behavior. Interactive technologies have been modeled on human cognition and perception. Centralized control systems, stimulus-response mechanisms and goal-oriented behaviors mirror the rationalist, linear paradigms of Enlightenment thought. This legacy has shaped the early development of computational interfaces and continues influencing designers' thinking on usability, feedback and affordance (Norman, 1988).

Even as design started to incorporate living materials, anthropocentric assumptions persisted (Bennett, 2010; Armstrong, 2012). In many early experiments, living organisms were used instrumentally without acknowledging their biological autonomy. However, biological systems do not easily conform to human expectations. They operate on different timescales, require specific conditions and cannot be fully predicted. Their "responses" are not always immediate or programmable. Consequently, the traditional design concept of control, the capacity to anticipate and engineer particular results, becomes impracticable in DILS. Instead, designers must embrace uncertainty, imperfection and co-agency (Dunne & Raby, 2013; Đan Vy Vu et al., 2024).

Consider, for instance, the difference between a microcontroller and a fungal network. While the former processes data using binary logic, the latter transmits nutrients, chemical signals and electrical impulses via dynamic, adaptive pathways. Fungi respond to changes in moisture, pressure and the presence of other microbes not through code but through a decentralized, embodied intelligence that challenges designers to think beyond linear causality (Adamatzky et al., 2022; Mayne et al., 2023). Furthermore, anthropocentric design often treats living organisms as static "features" of a product, ignoring their ability to grow, adapt or deteriorate. In contrast, DILS invites the possibility of artifacts that change over time, may outlive or outgrow their original functions and demand care rather than control. In such systems, the designer's role shifts from creator to caretaker and from problem-solver to cohabitant (Liu et al., 2018).

This redefinition has methodological implications. Rather than user-centered research focused solely on human needs and goals, DILS requires multispecies ethnographies, ecological modeling and biocompatibility assessments. The "user journey" may encompass human actions, microbial lifecycles, fungal metabolism and plant photosensitivity. Interaction maps must be redrawn to include touch zones and touchpoints, ecological thresholds where life meets interface. While foundational to the history of interaction design, anthropocentric design is insufficient for the complexity of DILS. Moving beyond it requires new tools and new ways of thinking (Biggs et al., 2024; Yanlic & Coskun, 2024).

Learning from Plants and Microbes

In order to reorient design toward regenerative perspectives, it is crucial to adopt ecological thinking. This is a systemic and relational method that recognizes the interdependence between humans, non-human organisms and their environments (Biggs et al., 2021; Wang & Pasquero, 2024). Ecological thinking moves beyond utilitarian and anthropocentric views of nature, promoting coexistence through care, reciprocity and co-evolution. Within this framework, plants and microorganisms are not just resources to be exploited but rather epistemological guides and co-designers. They offer sophisticated models of distributed intelligence, non-neuronal communication, ecological resilience and adaptive behavior. They propose alternative, relational and non-hierarchical approaches to interaction that are highly adaptive' (Ingold, 2013; Mancuso, 2017; Catellani et al., 2023; Calvo et al., 2020).

Plants, for example, do not have a central brain. Instead, they process information throughout their tissues, responding to light, gravity, touch and chemical gradients. As Stefano Mancuso (2017) argues, plant intelligence lies in their ability to perceive and react to their environment without centralized control. Plants grow toward light, avoid toxins and establish symbiotic relationships with fungi, all through decentralized signaling networks. This view is consistent with the idea that plant behavior is a form of minimal cognition enabled by root-based signaling systems (Calvo & Keijzer, 2011). This model contrasts significantly with traditional computational architectures, which rely on centralized processing, linear causality and fixed hierarchies.

In contrast, it closely aligns with DILS principles, as both operate through distributed, adaptive and context-sensitive mechanisms. In DILS, as in plants, information processing is not separate from physical form; sensing, responding and growing are deeply intertwined. Rather than executing preprogrammed instructions, these systems evolve in direct response to environmental stimuli, exhibiting emergent behaviors that defy prediction and resist centralized control. Just as plants rely on decentralized physiological networks to integrate light, moisture, chemical signals and social cues from other species (Calvo & Keijzer, 2011; Mancuso, 2017), DILS systems embody an intelligence that emerges through their ongoing interaction with their environment.

Similarly, microbial communities, such as bacterial biofilms or yeast colonies, display collective behaviors that challenge individualistic and reductionist models. Through processes such as quorum sensing, bacteria coordinate gene expression based on population density, resulting in emergent patterns of activity (Bassler, 2002; Waters & Bassler, 2005). These systems function through local rules, feedback loops and environmental adaptation, exemplifying forms of intelligence that are spatially distributed, relational and adaptive. As Nadell, Xavier and Foster (2009) explain, the sociobiology of biofilms reveals complex cooperative behaviors structured by internal communication and external environmental stimuli.

Learning from plants and microbes encourages a shift in our perception of materials. Instead of viewing materials as passive substances, we start to recognize them as living, responsive and dynamic entities. The concept of "material performance" thus encompasses biological growth, decay and regeneration. For instance, engineered living materials (ELMs) integrate living cells within polymeric matrices, creating substances that can grow, sense and adapt like biological organisms (Liu et al., 2022). Therefore, a moss-covered interface does not merely display information; it embodies and responds to conditions such as humidity, light and microbial presence.

Designers working with these organisms must familiarize themselves with the specifics of biological care. What are the needs of the bacteria embedded in a biointerface? How does the growth cycle of a lichen affect the temporal logic of interaction? What does it mean to maintain a system that resists optimization, grows unpredictably and may never "perform" in a conventional sense? These questions are not just technical; they are also philosophical and ethical. They compel us to rethink design as mastery over form and function and participation in the complexity of living systems (Camere & Karana, 2018).

POST-ANTHROPOCENE AND SYMBIOTIC FUTURES IN DILS

Redefining the Design Landscape

The term "Anthropocene", popularized by Crutzen and Stoermer in 2000, is still used "informally" to describe our current geological epoch, which is characterized by irreversible human impact on Earth's systems. In the context of climate disruption, biodiversity loss and exhaustion of resources, design is no longer merely a problem-solving or form-giving discipline; it becomes a means of shaping worlds. In the post-Anthropocene, design must shift its focus from extraction to regeneration and from control to cohabitation (Forlano, 2017; Davidová et al., 2020; Rognoli et al., 2021).

The shift from the Anthropocene to the post-Anthropocene era requires us to reconsider the role and responsibility of design. Rather than focusing solely on satisfying human needs or optimizing efficiency, design is now required to contribute to social, ecological and technological systemic regenerations. This transformation requires new skills and dispositions, such as an openness to uncertainty and a willingness to embrace slowness, share control and give up dominance (Karana et al., 2023).

DILS exemplify this emerging paradigm. Rather than treating living organisms as passive materials, DILS integrate them as active agents – co-performers, collaborators and, in some cases, users. These systems challenge the anthropocentric assumptions that have historically shaped interactive technologies, necessitating a re-evaluation of their purpose and ethics.

In this new landscape, designers have become biodesigners who are more like facilitators, ecologists or mediators than authors. Rather than imposing predetermined forms, they cultivate the conditions for life to emerge, thrive and interact. This involves observing fungal growth, attuning to microbial rhythms and designing not just for interaction but also for interaction within ecosystems. DILS are not designed for scalability or predictability but for vitality, care and entanglement with other species. They embody a shift from mastery to cohabitation and functionalism to relationality, paving the way for post-anthropocentric design futures (Pollini, 2023).

Symbiotic Design Models

As articulated by the evolutionary biologist Lynn Margulis (1971), the concept of symbiosis provides a compelling framework for reconsidering design in the context of DILS. Challenging the Darwinian view of evolution being driven primarily by competition, Margulis introduced the theory of *symbiogenesis*, which suggests that complex life forms emerged from cooperative relationships between distinct organisms. Mitochondria and chloroplasts, once autonomous bacteria, are now integral to eukaryotic cells, reminding us that life as we know it would not exist without symbiosis.

Design can also evolve by embracing this biological principle. Rather than exploiting living systems as resources or instruments, symbiotic design treats them as co-creators. This shifts the design focus away from extraction and optimization toward reciprocity, care and relational richness.

DILS exemplify this ethos in multiple ways. In fungal-based sensors, for instance, mycelial networks are not simply exploited for their conductive properties but are also recognized for their ecological intelligence. Similarly, moss-based responsive surfaces use humidity not as a controllable data point but as a condition of shared existence, an index of cohabitation. Concepts such as "thinking mycorrhizally" (Choy et al., 2009) provide a framework for viewing design as a distributed, collaborative system based on mutual benefit and exchange, like the underground networks connecting fungi and plant roots.

When applied to real-world contexts, symbiotic design can result in living infrastructures that purify the air, regulate microclimates and break down pollutants while simultaneously providing habitats for non-human life. Interfaces can be developed to respond to human input and modulate behavior concerning microbial activity, soil composition, moisture levels and light conditions. These systems represent a shift in design philosophy, moving from designing for nature to designing with nature.

Crucially, symbiotic design requires designers to work across multiple scales, from bacterial metabolism to ecosystem resilience, through a systems thinking approach that embraces feedback loops, ecological thresholds and emergent behavior. This mindset is attuned to complexity, adaptation and interdependence.

Symbiotic design delivers more than just functional or aesthetic innovation by integrating these principles. It redefines design philosophically and ethically as an ecological practice of cohabitation, where vitality, reciprocity and care become the driving forces of interaction and creativity.

POSTHUMAN AND MORE-THAN-HUMAN DESIGN

Decentering the Human and Designing with Entanglements

Decentering the human does not mean eliminating human presence but rather repositioning it as one actor among many within a relational meshwork. This approach requires us to acknowledge that the needs of a bacterial colony or a fungal network may be just as crucial to the success of a system as those of a human user, a perspective also central to recent work on more-than-human design practices (Nicenboim et al., 2023).

As Giaccardi and Redström (2020) emphasize, artifacts are never isolated; they are always embedded within larger socio-technical and ecological networks. They underline the *entangled* nature of contemporary design practice. DILS exemplify this entanglement, necessitating designers to consider not only interface aesthetics and usability flows but also energy cycles, maintenance rhythms and interspecies compatibility. Consequently, interaction design becomes a form of *world-making*, an act with ecological, ethical and political implications.

In this expanded design space, conventional methods such as wireframing and usability testing are enhanced through soil analysis, environmental simulation and long-term ecological observation. The design process becomes slower, more speculative and more iterative. This approach aligns with Anne Galloway's concept of "designing with care", where attentiveness replaces intervention and organic growth takes precedence over rigid execution.

Projects such as *The Plant Radio* (Rolighed et al., 2022) and *Living Bits* (Pataranutaporn et al., 2020) exemplify this emerging paradigm. Rather than seeking to simplify or commodify plant behavior, these works create interfaces for cross-species communication, translating bioelectrical activity into sound or digital data. These systems function not as conventional interfaces but as relational ecologies shaped by light, moisture, microbial presence and vegetal response.

Designing with entanglements, in this sense, means recognizing that technology is not an end goal but a facilitator of co-presence. This involves creating adaptive systems that decay gracefully and welcome participation from organisms whose lifeworlds we are only beginning to understand.

Ethical Implications of DILS

DILS blur the boundaries between material and organism and between designer and caretaker. This raises urgent ethical questions. What responsibilities do we take on when incorporating living systems into our designs? How can we engage with synthetic biology, biofabrication or genetic engineering without perpetuating the logic of extraction or instrumentalization?

A central issue is that of agency. Traditional design treats materials as passive and malleable toward a predetermined function. However, living systems have their own rhythms, needs and capacity for action. Designing with life necessitates a shift in mindset that recognizes the partial autonomy of biological entities. Another critical concern is care. Living materials require ongoing attention, nourishment and suitable environments. This challenges the notion of a finished product, instead emphasizing design's temporal and evolving nature. A DILS is not a static artifact but a living, dynamic system in constant transformation.

Synthetic biology magnifies these issues in particular. Genetically modified organisms (GMOs) have been used to produce biofuels, sense toxins or change color in response to stimuli. While such interventions offer potential benefits, they also risk commodifying life, turning living systems into programmable platforms.

As the designer and researcher Maurizio Montalti has noted, synthetic biology is a powerful tool that can be used for beneficial or destructive purposes (Albergati, 2022). For this reason, it must be subject to robust ethical regulation (Armstrong, 2022; Zohu et al., 2022; Kim et al., 2022). Transparent regulation, public discourse and inclusive ethical frameworks are essential.

In turn, designers must cultivate new forms of ethical literacy, drawing on environmental justice, indigenous epistemologies and critical bioethics. DILS represent ethical provocations as well as technical possibilities. By embedding life into artifacts, they encourage us to slow down, question our assumptions and consider alternative futures. Ultimately, they advocate a transition from user-centered design to *more-than-human–centred design*, where interdependence, care and respect for life guide creative practice.

CONCLUSION

DILS are not just technological innovations; they also raise philosophical and ethical questions. They compel us to reconsider what it means to design, interact and inhabit a world characterized by profound ecological interdependence. Rather than offering mere functional solutions, DILS challenge the assumptions of modernist, human-centered design, inviting a shift from extraction to regeneration, control to cohabitation and product to process.

In this emerging paradigm, design becomes an act of hosting rather than shaping, an attunement to multiple temporalities, metabolic logics and interspecies relationships. DILS are systems that are not only designed but also designed back. They demand that we cultivate a new design literacy, one that is grounded in ecological ethics, material agency and multispecies care – a literacy actively explored by Nicenboim et al. (2023) through the integration of posthuman theory and more-than-human design. Artifacts are no longer inert tools; they are living, dynamic systems with their own rhythms, needs and agency. The "user" is not necessarily human – it could be a bacterium, a fungus or a moss colony, for example.

This shift requires us to abandon the illusion of total control and embrace uncertainty, slowness and emergence. DILS defy predictability: they grow, decay, adapt and resist optimization. In doing so, they encourage designers to adopt a post-anthropocentric ethos that recognizes the complexity and fragility of life.

As Elisabet Sahtouris (2015) suggests, humanity may be entering a stage of evolutionary maturity characterized not by competition and consumption but by cooperation and restraint. DILS can serve as exemplars of this transition. They teach us to design not in isolation, but in relation; not for dominance, but for interdependence; not for the market, but for the web of life.

For design to play a meaningful role in the post-Anthropocene era, it must evolve materially, ethically and epistemologically. This means learning from fungi, collaborating with bacteria and caring for plants, not just metaphorically but materially. DILS illuminate this path forward. They embody a regenerative design practice that restores relationships, honors complexity and fosters mutual flourishing in the shadow of the Anthropocene.

REFERENCES

Adamatzky, A., Jones, J., Schubert, T., & Levin, M. (2022). Logics in fungal mycelium networks. *Logica Universalis*, *16*, 655–669.

Albergati, E. (2022). A conversation with Maurizio Montalti. Retrieved March 10, 2022, from https://www.pok.polimi.it/course/view.php?id=170#tab2

Armstrong, R. (2012). *Living architecture: How synthetic biology can remake our cities and reshape our lives.* TED Books.

Armstrong, R. (2022, June 25–July 3). Biodesign for a culture of life: Of microbes, ethics, and design. In D. Lockton, S. Lenzi, P. Hekkert, A. Oak, J. Sádaba, & P. Lloyd (Eds.), *DRS2022: Bilbao*. Design Research Society. https://doi.org/10.21606/drs.2022.144

Bassler, B. L. (2002). Small talk: Cell-to-cell communication in bacteria. *Cell*, *109*(4), 421–424. https://doi.org/10.1016/S0092-8674(02)00749-3

Bennett, J. (2010). *Vibrant matter: A political ecology of things*. Duke University Press.

Biggs, H. R., Bardzell, J., & Bardzell, S. (2021, May). Watching myself watching birds: Abjection, ecological thinking, and posthuman design. In *Proceedings of the 2021 CHI conference on human factors in computing systems* (pp. 1–16). ACM.

Biggs, H. R., Rosén, A. P., Veselova, E., Sánchez Ruano, D., & Cerna, K. (2024). Multispecies ethnography in design research and practice. In A. P. Rosén, A. Salovaara, A. Botero, & M. L. J. Søndergaard (Eds.), *More-than-human design in practice* (pp. 119–131). Routledge.

Braidotti, R. (2013). *The posthuman*. Polity Press.

Calvo, P., Gagliano, M., Souza, G. M., & Trewavas, A. (2020). Plants are intelligent, here's how. *Annals of Botany*, *125*(1), 11–28. https://doi.org/10.1093/aob/mcz155

Calvo, P., & Keijzer, F. (2011). Plants: Adaptive behavior, root-brains, and minimal cognition. *Adaptive Behavior*, *19*(3), 155–171. https://doi.org/10.1177/1059712311409446

Camere, S., & Karana, E. (2018). Fabricating materials from living organisms: An emerging design practice. *Journal of Cleaner Production*, *186*, 570–584.

Catellani, A., Cholet, C., & Pascual Espuny, C. (2023). The intelligence of plants: A new narrative? Notes on a contemporary cultural phenomenon. https://hal.archives-ouvertes.fr/hal-04183214

Choy, T. K., Faier, L., Hathaway, M. J., Inoue, M., Satsuka, S., & Tsing, A. (2009). A new form of collaboration in cultural anthropology: Matsutake worlds. *American Ethnologist*, *36*(2), 380–403. https://doi.org/10.1111/j.1548-1425.2009.01141.x

Đan Vy Vu, Funk, M., Huang, Y.-C. & Barati, B. (2024). Addressing uncertainty in biodesign through digital twins: A case of biofabrication with mycelium. *ACM Transactions on Computer-Human Interaction*, *31*(6), Article 69. https://doi.org/10.1145/3685271

Davidová, M., & Zavoleas, Y. (2020). Post-anthropocene: The design after the human-centered design age. In D. Holzer, W. Nakapan, A. Globa, & I. Koh (Eds.), *CAADRIA 2020: Re:Anthropocene – Design in the age of humans* (Vol. 2, pp. 203–212). Association for Computer Aided Architectural Design in Asia.

Dunne, A., & Raby, F. (2013). *Speculative everything*. MIT Press.

Escobar, A. (2018). *Designs for the pluriverse: Radical interdependence, autonomy, and the making of worlds*. Duke University Press.

Forlano, L. (2017). Posthumanism and design. *She Ji: The Journal of Design, Economics, and Innovation*, *3*(1), 16–29. https://doi.org/10.1016/j.sheji.2017.08.001
Giaccardi, E., & Redström, J. (2020). Technology and more-than-human design. *Design Issues*, *36*(4), 33–44.
Haraway, D. J. (2016). *Staying with the trouble: Making Kin in the Chthulucene*. Duke University Press.
Ingold, T. (2013). *Making: Anthropology, archaeology, art and architecture*. Routledge.
Karana, E., Barati, B., Giaccardi, E., & Nimkulrat, N. (2020). Living (with) things: A framework for designing interdependent materials and products. *International Journal of Design*, *14*(2), 101–117.
Karana, E., McQuillan, H., Rognoli, V., & Giaccardi, E. (2023). Living artefacts for regenerative ecologies. *Research Directions: Biotechnology Design*, *1*, e16. https://doi.org/10.1017/btd.2023.10
Kim, R., Zhou, J., Groutars, E. G., & Karana, E. (2022, June 27–July 3). Designing living artefacts: Opportunities and challenges for biodesign. In D. Lockton, S. Lenzi, P. Hekkert, A. Oak, J. Sádaba, & P. Lloyd (Eds.), *DRS2022: Bilbao* (pp. 1–10). Design Research Society. https://doi.org/10.21606/drs.2022.942
Liu, A. P., Li, X., Chen, Y., & Smith, J. (2022). The living interface between synthetic biology and biomaterial design. *Nature Materials*, *21*(4), 390–397. https://doi.org/10.1038/s41563-022-01231-3
Liu, J., Byrne, D., & Devendorf, L. (2018). Design for collaborative survival: An inquiry into human-fungi relationships. Unpublished manuscript.
Mancuso, S. (2017). *The revolutionary genius of plants*. Atria Books.
Margulis, L. (1971). Symbiosis and evolution. *Scientific American*, *225*(2), 48–61.
Mayne, R., Smith, J., & Lee, A. (2023). *Propagation of electrical signals by fungi*. arXiv preprint arXiv:2304.10675.
Morton, T. (2013). *Hyperobjects: Philosophy and ecology after the end of the world*. University of Minnesota Press.
Nadell, C. D., Xavier, J. B., & Foster, K. R. (2009). The sociobiology of biofilms. FEMS *Microbiology Reviews*, *33*(1), 206–224. https://doi.org/10.1111/j.1574-6976.2008.00150.x
Nicenboim, I., Oogjes, D., Biggs, H., & Nam, S. (2023). Decentering through design: Bridging posthuman theory with more-than-human design practices. *Human–Computer Interaction*, *40*(1–4), 195–220. https://doi.org/10.1080/07370024.2023.2283535
Norman, D. A. (1988). *The psychology of everyday things*. Basic Books.
Odenthal, F., & Suner, M. (2017, December 29). The anthropocene: Age of genocide. *FairPlanet*. https://www.fairplanet.org/developing-stories/saving-rhinoceros/the-anthropocene-age-of-genocide/
Pataranutaporn, P., Vujic, A., Kong, D. S., Maes, P., & Sra, M. (2020). Living bits: Opportunities and challenges for integrating living

microorganisms in human-computer interaction. In *Proceedings of the augmented humans international conference* (pp. 1–12). Association for Computing Machinery. https://doi.org/10.1145/3384657.3384783

Pollini, B. (2023). *Healing materialities from a biodesign perspective: Framing biodesigned materials and artefacts for the sustainability transition* [PhD thesis, Politecnico di Milano].

Puig de la Bellacasa, M. (2017). *Matters of care: Speculative ethics in more than human worlds.* University of Minnesota Press.

Rolighed, M. L., Aagaard, E. M., Due Jensen, M., Frankjaer, R., & Hansen, L. K. (2022). Plant radio: Tuning in to plants by combining posthumanism and design. In *DIS '22: Designing interactive systems conference* (pp. 666–676). Association for Computing Machinery. https://doi.org/10.1145/3532106.3533517

Rognoli, V., Pollini, B., & Alessandrini, L. (2021). Design materials for the transition toward post-Anthropocene. In B. Camocini & F. Vergani (Eds.), *From human-centered to more-than-human design: Exploring the transition* (pp. 101–130). Franco Angeli.

Sahtouris, E. (2015). The brink of disaster ... Or the brink of maturity? *World Future Review*, 7(3), 35–45. https://doi.org/10.1177/1946756714553249

Sánchez Ruano, D. (2025). More-than-human connections: Regenerating through multispecies design. In C. Cobreros, E. Giorgi, & T. Cattaneo (Eds.), *Regenerative design* (pp. 39–68). Springer. https://doi.org/10.1007/978-3-031-76890-3_3

Tironi, M., Chilet, M., Ureta Marín, C., & Hermansen, P. (Eds.). (2023). *Design for more-than-human futures: Towards post-anthropocentric worlding.* Routledge.

Tsing, A. L. (2015). *The mushroom at the end of the world: On the possibility of life in capitalist ruins.* Princeton University Press.

Wang, X., & Pasquero, C. (2024). Ecological thinking in regenerative architecture: Relevance of abduction in ecoLogic Studio's Deep Green research project. *Journal of Chinese Architecture and Urbanism*, 6(1), 1084.

Waters, C. M., & Bassler, B. L. (2005). Quorum sensing: Cell-to-cell communication in bacteria. *Annual Review of Cell and Developmental Biology, 21*, 319–346. https://doi.org/10.1146/annurev.cellbio.21.012704.131001

Winograd, T., & Flores, F. (1986). *Understanding computers and cognition: A new foundation for design.* Ablex Publishing.

Yanlic, B. S., & Coskun, A. (2024). Understanding the needs of nonhuman stakeholders in design process: An overview of and reflection on methods. arXiv preprint arXiv:2407.14750.

Zhou, J., Barati, B., Giaccardi, E., & Karana, E. (2022). Habitabilities of living artefacts: A taxonomy of digital tools for biodesign. *International Journal of Design*, *16*(2), 57.

INDEX

For Product Safety Concerns and Information please contact our EU representative GPSR@taylorandfrancis.com Taylor & Francis Verlag GmbH, Kaufingerstraße 24, 80331 München, Germany

Batch number: 10722108

Printed by Printforce, the Netherlands